MEMORY AND HEALING

MEMORY AND HEALING
Neurocognitive and Psychodynamic Perspectives on How Patients and Psychotherapists Remember

Soren R. Ekstrom

Routledge
Taylor & Francis Group

LONDON AND NEW YORK

First published 2014 by Karnac Book Ltd.

Published 2018 by Routledge
2 Park Square, Milton Park, Abingdon, Oxon OX14 4RN
711 Third Avenue, New York, NY 10017, USA

Routledge is an imprint of the Taylor & Francis Group, an informa business

British Library Cataloguing in Publication Data

A C.I.P. for this book is available from the British Library

ISBN-13: 9781782200321 (pbk)

Typeset by V Publishing Solutions Pvt Ltd., Chennai, India

To my sons,
Arne and Max

CONTENTS

ACKNOWLEDGEMENTS

Without the help of colleagues, friends, and family, this book would not have been possible. Throughout my work on this book, I have had the privilege of input from many of my colleagues. I am particularly grateful to Joseph Cambray, Thomas Putnam, and Pamela Newton, members of the Jungian community in Boston. I am also grateful to Jonathan Kolb of the Boston Psychoanalytic Institute for his insightful comments and suggestions.

For the entire process of writing of this book, I have had the true support of my family. My son, Arne Ekstrom, professor of neuroscience at the University of California at Davis, and my daughter-in-law, Eve Isham, also of the University of California at Davis, have given tireless attention to the referenced research and invaluable suggestions about studies that apply to my proposed integrations. My other son, Max, entrepreneur and computer programmer, has been a steady support by reading many of the early drafts and offering his encouragement. His wife, Keeley Schell, through her doctoral work in classics, offered many insights into mythological amplifications to dreams.

From the beginning, my wife, Carlyn, has been a thoughtful, perceptive, and invaluable reader. Her insights about creativity and psychological processes often offered additional perspectives on many of the issues in the book. She kindly understood why for hours I had to disappear into my study. And she knew how to pull me away from my desk for fresh air.

Carlyn's extensive readings and editing of the material were augmented by Leslie Cohen of CohenCopyediting, Cambridge, Massachusetts, and Leslie's skills as a professional editor. I could not have arrived at a finished product without her thoughtful corrections and comments. They have been critical in making my arguments clear and the text readable.

ABOUT THE AUTHOR

Soren R. Ekstrom, PhD, is a clinical psychologist and psychoanalyst in private practice in the Boston area. He has for many years been active in training future clinicians and has published papers about cognitive science and psychoanalysis in various journals. He has been the president of the C. G. Jung Institute of New York and served as an academic advisor for the Blanton-Peale Institute in New York City. Currently, he is an instructor and supervisor at the C. G. Jung Institute in Boston and one of the editors for the *Journal of Analytical Psychology*. His recent lectures have included such topics as narrative and metaphors in dreams and memory in the therapeutic alliance.

PREFACE

For some time, psychotherapists of all persuasions have been drawn to the rapidly expanding discoveries in neuroscientific research (Cozolino, 2002, 2006; Knox, 2003, 2011; Schore, 1994, 2003; Siegel, 1999, 2007, 2010a; Wilkinson, 2006, 2010). Partly, this may be due to frustration at psychotherapy's division into schools and paradigms, many of which by now rely on terms and theories that are over a hundred years old and thus no longer reflect current knowledge. Partly, the interest in neuroscience has been forced on the field by the success of neurochemical explanations for problems related to mental health. These explanations, although hardly conclusive, have coincided with the introduction of chemical treatments for a variety of mental disorders. They have also led to some unfortunate negative perceptions about the effectiveness of psychotherapy.[1]

This book explores why neuroscience—in particular, research on memory—offers new ways to understand why psychotherapeutic treatment is often an effective approach to a wide range of psychological problems (see Chapters Two and Eight). However, this book suggests that, to document that this is the case, we need to reformulate many terms and theories for what happens in treatments.

So far, the findings in neuroscientific research have mostly been introduced to clinicians using a general and neurobiological approach. The emphasis has been on identifying brain sites that may be involved in psychotherapeutic processes.[2] What is missing in this approach is a more direct integration. This book attempts to pave the way for such integration by showing how clinicians function in treatments and how, by extension, patients process and integrate new and transformative information.

Particularly relevant in this regard are findings on memory, a long-neglected function in formulations about psychotherapeutic approaches (see Chapter Eight). Discoveries about episodic and autobiographical memory, in particular, have a direct bearing on what happens in psychotherapy and will demand new approaches to clinical education and how clinicians report on their work. To document what happens to therapist and patient in a treatment, we need to know how they remember and how their experiences are accessed and integrated—thus, why the memory of both participants is intimately tied to a patient's experience of healing.

The neurocognitive processes activated in psychotherapy can only be fully explored when neuroscientists are enlisted in designing studies that directly measure the effects of the therapy process on both patients and therapists. So far, this is still to be done, and there are many legal and practical obstacles to overcome before it can happen. But psychotherapists also have to find ways to formulate what happens with their patients in terms that are useful to researchers. It is my hope that this book will provide some useful models toward that end.

A brief outline: Part I

The book has two parts. The first, "Applying the findings in research", is devoted to how current research applies to psychotherapy.

The first chapter, "Why memory and psychotherapy", presents the reasons why research on memory is critical to the practice of psychotherapy. Two neurocognitive models that have emerged from this research are outlined and contrasted with traditional psychotherapy formulations (LeDoux, 2002; Schacter, 1996). The chapter also references research on the connection between memory and narratives, and defines the memory function in psychotherapy based on the neurocognitive models (Lieblich, McAdams, & Josselson, 2004; McAdams, 2006).

Chapter Two, "The nature of subjectivity", deals with findings about conscious or explicit memory, including the fact that conscious memory is no longer regarded as unitary but as consisting of at least two different systems. One of these systems, episodic memory, is of great significance to psychotherapy and relates to first-person knowledge, or subjectivity. Contrary to everyday wisdom, specific methods now exist for studying first-person knowledge and much has been learned about how it is retrieved (Gardiner, 2002; Tulving, 2002).

Chapter Three, "Retrieving history of the self", examines research on autobiographical memory, a uniquely different kind of memory system that traces a person's history and sense of self. The chapter discusses a protocol developed by a group of British researchers that can be used to access autobiographical memory in a patient (Conway & Fthenaki, 2000).

Chapter Four, "Stories told and retold", explores the role of narratives in the consolidation and retrieval of memory. My proposal is that all psychotherapeutic treatments have a narrative foundation. I base this proposal on several existing explorations of the specific narratives that are the cornerstone of most psychotherapy and, in particular, psychodynamic treatments. Two narratives—those that patients develop and those that therapists bring to the endeavour—are central to the discussion.

Chapter Five, "Dreams as stories", deals with dreams as a more unusual form of narrative. Of particular interest is how the storylines in dreams are similar to those in other narratives. This is illustrated by the example of an initial dream from a patient. In reviewing the neuroscientific research on dreams and sleep, I also propose a conceptual framework (Foulkes, 1999; Hobson, 1999; LeBerge, 1998; Solms, 1997). In particular, I deal with the issues presented when a patient reports a dream in a treatment.

Chapter Six, "Metaphors and meaning" (the final chapter of Part I), introduces findings in cognitive linguistics and developmental psychology about certain image schemas that are present in the young child by two years of age (Dodge & Lakoff, 2005; Lakoff & Johnson, 1999; Mandler, 2004, 2005). Based on this material, I propose that dream narratives originate in special image-structures that can be traced to perception and motor control. These structures later become part of a person's repertoire of metaphors and are reflected in dreams, fantasy, and regular speech. The discussion is illustrated by the examination of four dreams, two of which are from Freud and Jung, two from my patients.

Brief outline: Part II

The second part of the book, "Remembering, reporting, and teaching", examines how central areas of psychotherapy reporting and education are affected by the findings in neurocognitive and outcome research.

Chapter Seven, "Where it happens and how", looks at how a new understanding of memory—in particular, the therapist's process of remembering—calls for major changes in how psychotherapists present their work and how they develop the ability to use memory effectively. Six forms of memory are defined, based on current understanding of memory and findings in psychotherapy research (Strupp & Binder, 1984). The chapter also outlines how implicit and nonverbal communication fit into this picture and examines how the therapist's memory has been understood in the past.

Chapter Eight, "What there is to tell", uses the understanding developed throughout the book to suggest certain remedies. It outlines ways of reporting and accounting for what occurs in treatments and suggests how therapeutic skills should be taught when considering the kinds of memory therapists rely on.

Chapter Nine, "Listening in a different state of mind", returns the discussion to Freud's original formulation about his own use of memory as a therapist. His descriptions can now be examined based on recent discoveries about episodic memory and autonoetic awareness. Listening in a different state of mind, a state of "self-knowing awareness" or mindsight, appears to be at the centre of all effective psychotherapy (Siegel, 2010a; Wheeler, Stuss, & Tulving, 1997), as apparently confirmed by extensive research on psychotherapy (Lambert & Barley, 2003; Wampold, 2001). The therapeutic alliance, as a specific bond that develops between patient and therapist, has been found to be the most significant variable—not paradigmatic formulations of one form or another (Safran & Muran, 2000).

PART I

APPLYING THE FINDINGS IN RESEARCH

One of the most fascinating achievements of the human mind is the ability to mentally travel through time. It is somehow possible for a person to relive experiences by thinking back to previous situations and happenings in the past and to mentally project oneself into the anticipated future through imagination, daydreams, and fantasies. In the everyday world, the most common manifestation of this ability can be referred to as "remembering past happenings." Everyone knows what this phrase means and what it is like to reflect on personal experiences, past or future, that are not part of the present

—*Mark A. Wheeler, Donald T. Stuss, and Endel Tulving* (1997, p. 331)*

CHAPTER ONE

Why memory and psychotherapy

To begin with, I would like to describe briefly what happens to me as a therapist when I am with a patient. In this instance, the patient—I will call him Brian—enters my office for his weekly session. We exchange short greetings and sit down across from each other. He begins telling me how he is feeling, sometimes referencing our last session. Every so often, I nod. *Plain red shirt*, I notice. *Still the same overweight frame as last week.* However, Brian's facial expressions are definitely livelier today, and he appears less troubled.

As I listen to him begin the session, neither his diagnosis nor a specific treatment plan comes to mind. I remember certain details of his history, but only much later in the session and then as scenes or scripted stories. I have not memorised his treatment up to now, nor could I write a detailed history of it. I am simply responding to a relational event that has followed pretty much the same format since we first started meeting in my office. My mind is focused on the here-and-now and on cues that may tell me something about what is going on in Brian's mind. I have no conscious agenda, but I am in an entirely attentive frame of mind.

As the session progresses, some things I notice make only a fleeting impression, leaving no lasting traces in my memory. Other impressions stay with me for a minute or two. Many of the things he says, or that I say, will have escaped my memory by the time the session is over. Then there are those exchanges, and perceptions of mine that lead to those exchanges, that inform what I later bring to bear on our conversation. When I take short notes for myself after the session, such exchanges are usually what come to mind.

Most emotionally charged communications would belong to this category. So would any fundamental change in the way Brian relates to me. To a large extent, my experience of his session, and my responses, are formed contextually and are based on cues that inform my responses. My understanding of what is transpiring and my possible recollection of Brian's past history depend on these contextual cues.

Given how little I consciously retain from session to session, how can I, as a clinician, understand the depth of a patient's personality, history, and current suffering? What are my tools—absent a consistent, reliable memory of the session—that make it possible for Brian and other patients to feel that I understand them and that I am able to help them to heal, to become psychologically healthy? And how, in spite of all its shortcomings, does my memory of what occurred in the session play a role in this process?

The literature on psychotherapy contains few accounts, if any, of how clinicians remember. Descriptions of the internal processes that psychotherapists rely on for their day-to-day functioning are largely missing. Can we assume that there are forms of memory that are particularly relevant to analytic practitioners, perhaps to all psychotherapists? Can we pinpoint the kind of memory functions that are involved, psychodynamically or otherwise?

We can certainly point to variations in this regard. Each practitioner's personality, experience, and training are reflected in how he or she works. But surely we can also assume common features in how therapists put their memory to use. In this book, I argue that a more informed understanding of this process will illuminate in crucial ways how we retrieve and bring into play the information most essential to the patient's recovery.

Neuroscientific research on memory

The current models for human memory that have emerged from brain research over the last decades are based on data that were largely unknown when the original psychoanalytic formulations about psychotherapy were made. Our progress in understanding mind and brain is in many ways astonishing. Antonio Damasio of the University of Southern California, a major contributor in neuroscientific research, summarises this development in the following way:

> Over a brief period of time, roughly twenty years, there has been an explosive development of new theories regarding the brain and its workings, as well as powerful techniques that allow us to study the brain experimentally—from the level of the individual nerve cells and molecules those cells require in order to operate to the level of the brain's macrosystems.

> (1999, pp. ix–x)

Some of these findings have not only changed how researchers view the mind-body interaction. They have also made tools available for understanding what happens in psychotherapeutic treatments—tools that previously were missing and that now need to be used in relation to a wide range of phenomena encountered in treatments.

The picture of the human brain that emerges from the research is of billions of cells at work, as in a factory in twenty-four-hour–production mode. Whether we are fully awake or deeply asleep, lost in thought or staring at the wall, certain circuitries remain active. Chemicals are in perpetual transmission aided by biologically propagated electrical charges (LeDoux, 2002). The brain is constantly transmitting signals, which are turned into information—most of this going on implicitly, outside the reach of consciousness. Memory, like all other functions of the brain, such as perception and emotion, is not held in local storage, as if recorded by a camera or electronic device, but is the result of specific neurons working or firing together; a process also known as *Hebbian plasticity* (ibid., p. 79).[1]

Model one: the synaptic self

One of the models proposed by neuroscientists reflects this plasticity of the brain. Joseph LeDoux of New York University's Center for Neural Sciences, revisiting the problem of self and synaptic organisation, writes:

> Let's start with a fact: People don't come preassembled, but are glued together by life. And each time one of us is constructed, a different result occurs. One reason for this is that we all start out

with different genes; another is that we have different experiences. What is interesting about this formulation is not that nature and nurture both contribute to who we are, but that they actually speak the same language. They both ultimately achieve their mental and behavioral effects by shaping the synaptic organization of the brain. The particular patterns of synaptic connections in an individual's brain, and the information encoded by these connections, are the keys to who that person is.

(2002, p. 3)

What LeDoux refers to as "shaping the synaptic organization of the brain" is a process commonly called *synaptic plasticity* and is viewed as an "innate capacity for synapses to record and store information" (2002, pp. 9, 307).[2] Many psychological and behavioural functions central to psychotherapy theory are mediated by cells joined by synapses and working together. The trillions of synapses that allow the brain's cells and neural networks to communicate are themselves modified by experience, a process that represents learning (LeDoux, 1998, 2002).

When equating memory with learning, nature and nurture for LeDoux are no longer opposing forces. And the simple fact that we have genetic dispositions does not·mean that our experiences cannot interact with those dispositions and mould them. They do; and they do so by affecting the entire organisation of the brain. Memory reflects learning acquired by the brain and is the key to the entire organism's survival. Memory, a function only rarely mentioned in most psychotherapy formulations, becomes one of the most critical functions in our understanding of the human mind (Ekstrom, 2004; Schacter, 1996).

Also, according to this research, memory consists of both conscious and unconscious systems. Some memory may even be labeled *nonconscious* in that it functions automatically, beyond awareness, as when we apply grammar to our thought processes (LeDoux, 2002). The unconscious systems, often called *implicit*, are devoid of the subjective internal experience of correctly recalling information, such as our sense of the time or place for when an event occurred.

As Daniel Siegel (2003), of the UCLA School of Medicine, describes it, implicit memory manifests both behaviourally and emotionally. As *procedural memory*, it consists of skills like riding a bike that, once learned, do not require intentional retrieval (Schacter, Wagner, & Buckner, 2000). As *emotional memory*, it often retains painful experiences and makes us avoid situations that may have involved the cause of such pain, without our understanding why (Siegel, 2003). A third kind, particular to early developmental phases in children, is perceptual and event based and retains a sense of familiarity about perceptions and bodily experiences, but without the sense of remembering (Nelson, 1993).

Model two: engrams and mapping

Another model that has emerged from neurocognitive research is that of the *engram*. Using a term invented by Karl Lashley in the 1950s, Daniel Schacter, one of the leading neuroscientists focusing on memory, defines engrams as "the transient or enduring changes in our brains that result from encoding an experience" (1996, p. 58). When encoded in long-term memory, engrams are the brain's records of an event. This record consists of numerous sights, sounds, actions, and words, all being analysed in different sites of the brain, by different groups of neurons.

In the process, these neurons remain connected to one another. They form neural maps that link one part of the brain to another part of the brain. This process, also called *mapping*, does not produce a finished product. New patterns are constantly being formed, and most of this activity happens outside our awareness. However, when some of these patterns become conscious, they may be experienced as images (discussed in Chapters Five and Six, concerning dreams), or they will be encoded as memories (Damasio, 2010).

Most of the enduring engrams are conscious or explicit memories. However, findings in neuroscience do not support the long prevailing notion of a unitary conscious mind. As examined in detail in the next chapter, the knowledge that we rely on for our everyday lives is not simply of the factual and objective kind. We also have the capacity to remember uniquely subjective experiences and reflect on their meaning (Tulving, 2002).

Depth psychology and neuroscience

Psychodynamic formulations can be traced to an older explanatory model, depth psychology. According to this model, the mind or *psyche* consists, on the one hand, of the immediate self or the *ego* as the centre of consciousness; and, on the other hand, of unconscious complexes that may have a determining influence on the conscious mind but function outside it (Roazen, 1975).[3] As an abstraction, psyche is not specifically tied to memory; its functioning cannot be traced to neural networks or brain sites. Rather, the notion of psyche is rooted in a long Western philosophical tradition in which soul and matter are two separate qualities (Pinker, 2002). Unless seriously interfered with, consciousness, accordingly, is a distinct and unitary function.

As purely philosophical ideas, there were no direct ways to test the implications of depth psychological concepts in controlled studies. At the time the particular psychoanalytic versions were formulated, neuroscience did not exist and the little psychological research that was done was still the domain of psychiatry, also a fairly new discipline. The first to put depth psychological ideas into use were physicians concerned with a kind of suffering that seemed to have no obvious and discernable physical origin: patients, primarily women, were languishing in sanatoria, unable to live a meaningful life and tortured by the most horrendous memories (Ekstrom, 2004; Shorter, 1997).

Contemporary neuroscience approaches memory from different perspectives and has opened for review many established concepts about the workings of the mind. In some instances, these approaches add several new dimensions to the notions central to psychodynamic thinking. In others, they contradict them. From a neurocognitive perspective, the production of synaptic connections is part of the process of encoding information into a wide range of memory systems. As we have seen, these systems are understood—in much agreement with psychodynamic thinking—to be both a conscious and an unconscious kind. From a neurocognitive perspective, however, what we learn from experience, even when maintained in one of the unconscious or implicit memory systems, is stored for future reference. Thus, memory as a whole is future oriented. Even the one particular system that relates to past experiences has this orientation toward what will be of use in future situations (Tulving, 2000).

This neuroscientific perspective certainly adds a dimension to the depth psychological understanding in which memory is mainly a record of the past (Schafer, 1983).[4] In psychodynamic thinking, there is no real emphasis on learning. Experiences found unbearable or unpleasant may affect current behaviour, but new learning is generally ignored.

Emotion and memory

The role of emotion in conscious states is another area where the views in depth psychology and neuroscience diverge. A common understanding among clinicians is that strong emotions or affects interfere with consciousness and overwhelm a person's attention, even cause dissociation and cognitive lapses (Wilkinson, 2006).

For neuroscientists, emotion has an equally important role in how memory is reinforced. What James McGaugh of the University of California at Irvine calls *emotional arousal* enhances memory "by activating systems involved in regulating the storage of newly acquired information" (1995, p. 256). Even when certain experiences are repressed and dissociated from consciousness, as depth psychology maintains, emotions consolidate what is being remembered so that it can be maintained for the long term (Wilkinson, 2004). Emotional arousal may therefore consolidate memories, not merely repress them.

To what extent such consolidation affects conscious memory or causes dissociation and fragmentation by being implicitly encoded is still unclear. For LeDoux (1995), subjective and mind-changing emotional states in which memory is consolidated depend on the robustness of the person's consciousness. Under favourable conditions, such subjective states tend to "dominate consciousness until they dissipate" (p. 1059). However, it is also true that fear, one of the emotions most studied by neuroscientists, may in some instances activate powerful defences. Instead of producing mind-changing states, self-protection activates powerful survival mechanisms, a sign of trauma responses (Van der Kolk, 2003).

The therapist's learning is no exception in this regard. A healthy modicum of arousal is the optimal state of mind and allows for what LeDoux calls "subjective emotional states" to be maintained (1998). The absence of such emotional involvement, in the form of detachment, can no longer be seen as a desirable response (LeDoux, 1998; McGaugh, 1995). Yet, for certain psychoanalytic approaches, therapists train themselves to be in semi-trances, mindsets high in absorption but weakly encoded (Spieger, 1995). Therapists enter these mind-sets during especially intense experiences and when attempting to enter the patient's inner experience. According to Thomas Ogden of the Psychoanalytic Institute of Northern California, in such reveries the therapist "renders his own unconscious receptive to the unconscious of the analysand",

and thus sets aside certain conscious processing of experience (1997).[5] Consolidation of memory may therefore have to be compromised to accomplish this.

As David Spiegel of Stanford University Medical School points out, one of the characteristics of such reveries is an unusual degree of absorption and "an immersion in a central experience at the expense of contextual orientation" (Spiegel, 1995, p. 130). If not rehearsed immediately after the session in which it occurred, most of the information from such reveries will be lost. The contextual details may also be distorted, especially when the therapist attempts to account for the experience at a later date.

These research findings shed new light on what the therapeutic literature calls *induced countertransference reactions* (Safran & Muran, 2000). By seeing how therapists may enter a heightened state that dovetails with important memory functions, we have a better understanding of why they may find it difficult to account for their own involvement in what occurred in certain sessions. Particular phases of a therapy may, in fact, only be remembered as something that occurred to the patient alone. The way the therapeutic situation is constructed—as a repeated occurrence in the same place and with the same specific duration—may in fact facilitate this one-sided perspective.

When one considers the role of emotional arousal in how memory is encoded, stored, and retrieved, a negative view of the therapist's emotional involvement is misplaced, for several reasons. Emotion facilitates encoding of important information about the patient and makes memory of what occurred in a given session more durable. And it does so by the therapist's having access to what memory research calls *autonoetic awareness*. The next chapter looks at why autonoetic awareness, as a particular state of mind, is critical to remembering significant developments in a treatment from session to session.

The psychoanalytic view

It should now be clear that many of the theoretical formulations about psychotherapy, especially those with roots in psychoanalysis, no longer stand up to the findings in neuroscientific research. While psychodynamic formulations based on depth psychology may have made it possible to explore what at the time was new and uncharted territory, findings in neuroscience now require modifications and revisions of several common formulations about therapeutic treatments (BCPSG, 2010). Especially when it comes to the therapist's memory, these findings offer new and unique tools.

What constitutes the person's *self* must be incorporated with our knowledge of how the brain dictates principles of psychology— ironically, the approach Freud attempted to follow in his early and pre-analytic research (1895). Under normal operating circumstances, body, mind, and brain function together. They do so as manifestations of a single organism; and this organism produces many processes that operate nonverbally and internally, beyond our awareness (Damasio, 2003).

Even researchers friendly to psychoanalysis have pointed out that new neuroscientific findings must become part of how we understand psychotherapeutic processes. Mark Solms and Oliver Turnbull (2002), two cognitive researchers with analytic training, question the common claim among some therapists that Freud's discoveries can never be scientifically verified from the outside and only in the treatments themselves. To Solms and Turnbull, these therapists ignore the historical background: isolation from the academic community as a temporary strategy during a development phase of Freud's theories. Moreover, they argue, some followers of Freud have now "turned it [isolation] into an article of faith, and psychoanalysis has suffered as a result" (2002, p. 298).

One such area of faith is the assumption that previous experiences can be recalled more or less in the form they were first experienced. Although the patient's distortion of memory has been how therapists understand transference phenomena, they have also assumed that past trauma could be reclaimed without any factual distortions (Schacter, 1996). Whereas, memory, in today's neuroscientific understanding, is continuously updated and changed. Furthermore, it is also prone to decay. In fact, research on what is stored short term, as *working memory*, shows that major portions of what is temporarily remembered will soon be lost permanently (Baddeley, 2000).

Implicit or unconscious memory, in particular, represents learning that occurred via different routes and different neural pathways. It does not exist as a discrete part of the brain, as assumed in much of the early analytic thinking. The unconscious was, in fact, conceived statically, based on various theories, including drive theory, archetypes, and relational needs (Siegel, 1999). From a neurocognitive perspective, however, any recall of past events can never be a replication of experiences but is a complex mental reconstruction (Schacter, 1996).

Implications of the research

The implications of the new findings about memory are several. For one, patient data that psychotherapists rely on, almost exclusively, is information about the patient's experiences in the past. The patient's disclosures in the consulting room are verbal, narrative, and episodic; and this is the case whether the information is about the immediate past, such as what happened after the last session, or the remote past and significant childhood experiences. An understanding of the particular memory systems involved in this type of information is therefore imperative (see Chapters Two and Three).

The second implication has to do with the therapist's own memory. Our understanding of the therapeutic process must include the therapist's ongoing learning—his or her continuous integration of new information—as a significant element in treatments. This means, specifically, that theory and case reports can no longer focus on what happens exclusively to the patient. Neuroscientific research shows a direct link between memory and learning; in fact, the two are identical when considering a wide range of findings (LeDoux, 2002). How this learning takes place in the therapist, both consciously and unconsciously, is therefore of utmost importance, in terms of both how treatments are reported and how therapeutic skills are taught for training purposes. A significant portion of what the therapist learns is in response to the particular patient under treatment, apart from theoretical formulations and generalisations in the form of techniques.

All treatments must be regarded as dyadic processes—what Robert Stolorow and his colleagues call *intersubjective processes*—in which both parties influence each other's perceptions (Stolorow, Brandchaft, & Atwood, 1987). In this sense, therapy is never a simple dialogue between two participants but an interchange that happens verbally and nonverbally, explicitly and implicitly, involving both participants. At its core, it is based on a relationship that can only be understood in a two-person perspective (Wachtel, 2008).

However, in light of findings in memory research, the most significant element in a given treatment is how the therapist remembers and how this is communicated. We may never be able to document fully how learning takes place in a patient from session to session. But we can assume that the effectiveness of the therapist's communication is a major factor in such learning. Patients absorb and eventually integrate into their own life outside the therapeutic setting how the therapist uses his or her memory. Thus, they learn from how the therapist remembers. But what the therapist remembers will also activate further recollection in the patient. What the patient then remembers makes the therapist understand yet more aspects of the patient. In this reciprocal and dyadic process, the therapist learns from what the patient remembers, the patient from how the therapist remembers. Psychotherapy treatments must therefore be understood as intersubjective—as mutual activation and reactivation of memory.

In this light, *how therapists learn* may be the missing link in our understanding of what happens in treatments. The power of the process cannot be explained in any other way, especially if we consider how many of the detail from treatments is lost from session to session, if not within minutes or hours of a given session.

As my description of working with Brian shows, memory for everyday factual details thus becomes secondary. For therapists, the likelihood of losing much of what transpired in a session increases rapidly from the first hours of trying to reconstruct what occurred (Spence, 1982, p. 231). There is also a risk that what they report will be skewed by hindsight bias and that they will stick to their original assumptions about the therapeutic outcome, even when these assumptions are later contradicted by the patient data (Metcalfe, 2000).

Without accounting for how the therapist's memory (which I term *treatment memory*) functions, we leave out how the patient's recall of critical memories is cued and how the therapist's memory is put into use; and we are limited to describing certain interpersonal dynamics and inferred processes in the patient. These descriptions, without including the therapist's contribution, are inevitably incomplete. And when they are made long after the actual sessions, they become fictional reconstructions, even when based on the therapist's explicit notes after each session.

Unfortunately, therapists have grown accustomed to this one-sided way of accounting for therapeutic processes. Under the influence of medical and procedural thinking, they have felt competent to report how the patient experiences treatment, while rarely reporting on their own contributions. Such accounts typically begin with stating information about the patient's age, occupation, marital status, and sex. Without mentioning the relational context, they then proceed to the diagnosis before venturing into descriptions of the patient's responses and process of healing. And the only instances in which the therapeutic relationship is discussed are often cloaked in technical terms, as forms of transference and countertransference in response to that relationship (Messer & Wolitzky, 1997).

Treatment outcomes

Applying the new neuroscientific findings to how the therapist's memory works should also allow for a better understanding of what happens for the patient. Although we have no immediate ways to document the specifics of each patient's memory and learning, years of outcome studies provide us with some important answers as to what works and what types of outcomes can be expected (Wallerstein, 2001).

Michael Lambert and Dean Barley, two psychologists from Brigham Young University, recently reviewed data from a multitude of such studies (2002). Their review shows that psychotherapy is generally effective, with the average treated client better off than eighty per cent of untreated control subjects. In some studies, therapists with more than six years' experience seemed to have more success. In others, certain therapists seemed to have more success with difficult cases. But, overall, *how the therapist was perceived mattered the most* (Norcross, 2002).

The outcome research has now reached a point where the therapist— not methods and techniques—is coming into focus as the decisive factor in treatments. Many years of (often) large-scale projects have yielded no clear-cut proof about the superiority of one paradigm over another (Lambert & Barley, 2002). The tools being used by these researchers are also ideologically neutral. They focus on the dyadic nature of psychotherapy as a series of relational interactions (Miller, Luborsky, Barber, & Docherty, 1993; Strupp & Binder, 1984).

In describing the experiences of the Penn Project, a five-year psychotherapy research project, the team of Luborsky, Crit-Christoph, Mintz, and Auerbach conclude that no particular measure, before or after treatment, predicts outcome. However, they suggest that five factors are particularly relevant when looking at the outcomes in the entire project (1988, p. 271):

1. The patient's experience of a helping alliance
2. The therapist's ability to understand the patient
3. The patient's level of self-understanding
4. The patient's decrease in the pervasiveness of conflicts
5. The therapist's, as well as the patient's, ability to assist in internalising gains.

These suggestions imply that the individual therapist plays a significant role in treatments; and, further, that this is due much less to theoretical acumen or methodology than to conveying understanding, forming a helping alliance, and assisting the patient to integrate gains in understanding throughout the treatment process. The patient's level of self-understanding, to a large extent, has to do with the type of narrative that forms in treatments, a narrative that therapists play a critical role in developing. Only via such narratives can previously dissociated memory be fully integrated (Cozolino, 2002). In addition, such a narrative creates a cohesive new story from many disconnected ones, a *self-narrative* that for the patient can span a more extensive time period and encompass a fuller sense of self (see Chapter Four).

The therapist's role in the development of such self-narratives depends primarily on how competently the therapist remembers what occurs from session to session. This competence has several elements. Some originate in the therapist's personal therapy or training analysis, thus in the therapist's own self-narrative. This unique element in how psychodynamic therapists are trained—via a person-to-person transmission of skills—may explain the development of schools, but it does not, as I will discuss in Part II of the book, fully account for how other necessary memory components can be trained.

Treatment memory

For the purpose our discussion, I define the therapist's memory of a treatment, *the treatment memory*, as follows (see further, Chapter Eight):

a. It originates in the self-narrative therapists first develop in their own psychotherapy. These narrative structures are also used when understanding self-narratives by others (Shank, 1999).

b. It is first and foremost a type of expert memory that psychotherapists continue to develop in their practice, and it allows therapists to remember critical details from each session with a patient.

c. It is, therefore, the foundation upon which patients integrate their autobiographical knowledge, their own self-narrative. By providing possible cues to the patients' memories, it facilitates integration of otherwise disparate parts of patients' experiences.

As I will show in the following chapters, cognitively, treatment memory consists of several elements. These elements can be categorised using critical findings in neuroscientific research: on the one hand, explicit or conscious skills exist in the form of *semantic* and *episodic memory*; these skills are learned and continually updated. The basis for these conscious skills, most importantly, is the ability to process the episodic information encoded in each treatment. I term it *session-related memory*, a distinctly conscious skill, mainly activated by contextual cues.

On the other hand, therapists also rely on implicit, not fully conscious memory functions. These functions are primarily procedural and operate outside the therapist's awareness of origin and sources (Schacter, 1996). Some of the implicit memory functions consist of beliefs and predictions that develop from previous experiences. As condensed into mnemonic devices or *scripts*, these beliefs and predictions provide the therapist with hypothetical interpretations. Other memory functions— in particular, those based on procedural memory—are acquired by performing repeated tasks of a similar nature, from scripts of the general conduct of a session to certain listening skills (Schacter, Wagner, & Buckner, 2000).

We may find these and similar memory functions used by all professionals who, broadly speaking, counsel others. One function in particular is a form of narrative memory; I call it *interpretative memory*. It consists of a story-based understanding that therapists develop and continually update to reflect an understanding of common dynamics in treatments.

Its counterpart, *paradigmatic memory*, represents a theoretical framework that therapists use as an aid primarily when reporting entire treatments. In certain ways also a master narrative articulated as theories and concepts, paradigmatic memory is a *semantically based structure* that seems resistant to change (Schafer, 1980). Thus, it is more like an expository text and is usually unaffected by and unrelated to particular treatments. Although paradigmatic memory articulates a certain theory-based understanding of the patient's symptoms and personality, this type of memory is unable to hold information about what occurs with a patient from session to session.

Other memory functions that therapists rely on are associated with the therapeutic setting itself. They are accessed without much conscious effort and thus have to do with skills that develop from repetition. These elements—commonly called *basic techniques*—in a neurocognitive vocabulary are *procedural* and are the result of training, but mostly of repeated use (Solms & Turnbull, 2002). They form what I term *technical memory*. By training, therapists use these procedures to give the patient a sense of being protected and safe. They also give the therapist a sense of control of the therapy situation and their own responses. Everything from the greeting ritual to seating arrangements, to quiet listening instead of arguing for points, is something therapists learn from repeated actions.

Memory and narratives

In discussing how patients and therapists remember a treatment in later chapters, I also explore findings showing how narratives aid many memory functions. Stories, according to this research, are central to the way our experiences are organised and how we give them meaning. They are what Louis Cozolino of Pepperdine University, in his study of neuroscience and psychotherapy, calls "a vehicle of explaining behavior and defining both the social and the private selves" (2002, p. 166).

To Roger Schank of Northwestern University and others studying social cognition, much of what is needed to manage daily living is in fact remembered by creating a story in which several interrelated events can be included (Schank, 1990, 1999). By fitting our experiences into a story and making intelligent use of indexing, far more than simple event-based storage is possible. However, the more complex the data and the decisions we face, the more we also have to update our stories or create new ones. Schank writes:

> The process of story creation, of condensing an experience into a story-size chunk that can be told in a reasonable amount of time, is a process that makes the chunks smaller and smaller. Subsequent iterations of the same story tend to get smaller in the retelling as

more details are forgotten. Of course, they occasionally get larger when fictional details are added. Normally, after much retelling, we are left with exactly the details of the story that we have chosen to remember. In short, story creation is a memory process. As we tell a story, we are formulating the gist of the experience which we can recall whenever we create a story describing that experience.

(1990, p. 115)

In retelling stories, we are therefore engaging in a highly selective process of condensing and fine-tuning memory. In so doing, not only will the number of events remembered be much smaller, the original experiences now have coherence. This, in turn, makes reconstructing missing or loosely connected details easier than creating a new story each time we wish to relate a set of events. Even when fictional details and elaborations occur, the essence of the narrative remains.

Those who study narratives in literature echo these conclusions. According to Jonathan Culler, a scholar in comparative literature, by studying basic narratives we can demonstrate the centrality of literary structures to the organisation of experience. The logic of a story is something that supersedes the presentation of events.

The theory of narrative postulates the existence of a level of structure—what we generally call "plot"—independent of any particular language or representational medium. Unlike poetry, which gets lost in translation, plot can be preserved in translation from one language or one medium into another: a silent film or a comic strip can have the same plots as a short story.

(Culler, 1997, pp. 85–86)

Narratives, then, have an independent underpinning, also called *plot* or *storyline*, a story within the story. In presenting a series of events, the plot and the events are what Culler calls "a nondiscursive, nontextual given", something that exists prior to and independent of a verbal presentation, the discursive version (1981, p. 171). (The plot and the events will, as seen in the discussion of the origins of narratives in Chapter Four, be stored and retrieved differently.) A story, in other words, may be given a seemingly endless number of versions. The plot, on the other hand, is the narrative underpinning that allows the presentation to occur and gives it a particular meaning.

This makes the storyline especially significant as the part of a story we retain in memory, even if being used in several stories or versions of a story. In fact, storylines may have their origin in learning that occurred well before verbal articulation was possible (Mandler, 2004; Wilkinson, 2010). They provide a story with its structure and they do so by a sequence of actions over time, usually with a beginning statement, a turning point, and a resolution (Dautenhahn, 2002).

The multiplicity of storylines makes it clear that there are many ways to process what occurs in a treatment and that many versions are inevitable. Each account may relate many of the same events and experiences, but the narrative underpinning will probably differ. Thus, storylines are not stories as conceived in common parlance. Rather, they are the metaphorical element that makes each version of a series of events a "story". And they do so by conveying surprise, relief, or a sense of redemption (Mar 2004).

In presenting what occurred in a session, therapists might remember particular exchanges and disclosures. And when wishing to relate their relevance to an entire treatment, therapists need a storyline that gives a meaningful context for what happened. This storyline will most likely be found by referencing certain condensed stories, experiences that date back not only to the therapist's own personal therapy when in training, but also to his or her early childhood experiences and attachment pattern (Cozolino, 2002). Presenting what appears to be a straight account of a psychotherapy session involves more than remembering the particulars from the session.

Story vs. *narrative*

For a full story to emerge, it also needs to include what Robert Scholes of Brown University calls "sufficient continuity of subject matter" (1980, p. 206). Even when the story has relevance only to the therapist and the patient as the two participants in a treatment, it is being shaped by the implicit purpose of the treatment and how the treatment is being conducted. Although this story may place certain events in a sequence, that sequence is being fashioned by the subject matter. By appealing to a broader human drama, it elicits interest (Bruner, 1991). Scholes explains:

> When the telling provides this sequence with a certain kind of shape and a certain level of human interest, we are in the presence not merely of a narrative but of story. A story is a narrative with a certain very specific syntactic shape (beginning-middle-end or situation-transformation-situation) and with a subject matter which allows for or encourages the projection of human value upon this material.

> (1980, p. 206)

This distinction is not always maintained in the research on narratives and memory. However, it needs to be emphasised that well before therapists' narratives become stories in Scholes's sense—such as case reports that may resemble literature or fiction—they are mnemonic devices, part conscious, part unconscious, explicit as well as implicit.

Therapy accounts, or case reports, differ from fictional narratives in one important respect: they attempt to capture the relevance of actual events. Their purpose may not be to mirror exactly the events remembered, but they narrate the therapist's perceptions of actual events, at least at the time of formulating the report. Thus, the therapist's account of a certain treatment is an attempt to order, chronologically, statements and events that happened in that treatment. In this sense, therapists' narratives are nonfictional. They reflect choices as to what appeared most relevant and what elicited the individual therapist's emotional interest, but, importantly, they also string together statements and events for which the sources could be checked (Bruner, 1991; Scholes, 1980).

This is true also for stories by patients, although they rarely cover the entire treatment process. In telling their stories to a therapist, first and foremost, patients wish to elicit emotional interest. Their stories are nonfictional, as well, in that they also connect statements and events that probably could be verified and for which a high degree of believability exists (Culler, 1997). Similarly, when presenting their stories to others outside the therapy relationship, patients usually wish to create order in what they experienced by choosing what appeared most relevant in that particular setting.

We often associate plot and story with fiction, with literature and fine art. But historical accounts, legal and philosophical arguments, and case reports—just to mention a few—are all forms of narrative discourse and involve such known literary vehicles as plot, metaphor, argument, summation, and repetition. Only when we know that the events in a story are fictitious will their sources be irrelevant. Only when interpreting and evaluating the truthfulness of historical events may it be significant to bring extratextual information to bear on them and to ensure that what occurred will not be misrepresented or placed in artificial constructs. In a piece of fiction, however, the events themselves cannot be examined in this manner. They only exist in the text and for it. The story's veracity cannot be questioned or tested (Scholes, 1980).

As I will illustrate throughout this book, the meaning of the term *narrative*, as a nonfictional and historical account, has become an integral part of how psychotherapy is generally discussed. In particular, the term is used to denote a treatment narrative as a form of summarising a given therapy (Winer, 1994). Although this usage reveals problems of credibility in regard to case reporting, it has brought about an important debate about what happens in treatments and by whom the treatments can reliably be reported. Is it, in fact, possible for the therapist to objectively report on how the patient experienced the treatment? Or is the case report only to be regarded as an approximate or even hypothetical description of what *may* have happened?

Defining narratives

The term *narrative* thus has several meanings, such as:

1. *a sequential element* that makes it possible to show how a given series of events unfold over time, thus giving them an historic meaning (McAdams, 1993);
2. *a formal element*, also called *plot* or *storyline*, that organises the events to be related, usually as a particular topic or subject matter and according to certain causal principles;
3. *an emotional tone* meant to elicit interest and determined, or at least influenced, by the situation in which the story is told and the interaction of teller and audience.

In this book, I use case vignettes, both published and unpublished, to establish the nature of these continuously evolving structures as therapists use them. I use the term *narrative* to describe how therapists connect separate past events to each other and how these events are tracked "at least long enough in order to extract their portent for the future" (Westbury & Dennett, 2000, p. 13).

In his classical examination of historical accuracy in case reporting, Donald Spence of Rutgers Medical School discusses how contemporary psychoanalysis had to deal with two of Freud's legacies: on the one hand, his stated recommendation of maintaining unbiased attention; on the other hand, his written cases. "As analysts", Spence argues, "we may try to listen with evenly hovering attention, but we are also heavily, if unofficially, guided by the narrative tradition, and what registers may more often represent a good pattern than what precisely was said" (1982, p. 33).

What Spence regards as "a good pattern" also includes the well-rehearsed aids that therapists use to organise their experiences. These aids represent particular paradigms and are the result of the particular training and outlook of each therapist. These paradigms are, therefore, part of what I will call *paradigmatic memory*, learned master narratives onto which each treatment is attached as the therapist's own story of a treatment (Schafer, 1983). Robert Winer is another analytically trained therapist concerned with how case reports have been misunderstood. He concludes:

> In presenting the story of our work with a patient, we must always both organize the work and be selective, presenting one aspect and not another, and generally presenting the work as we eventually come to understand it, not as how we understood it at the time. But we are always organizing our experience with our patients, from the first moment we see them. And that organizing is always as much about us—us as the analyst, us as the person—as it is about them. Case reports are never, in fact, about patients. They are always about treatments, about our view of an interaction in which we have participated, no matter whether we have chosen to recognize or to conceal it.
>
> (1994, p. 111)

To Winer, a case report is always based on hindsight. Details that seemed relevant at a certain time in a treatment tend to get lost, while the overall understanding is editorialised and fine-tuned. A case report is a narrative as much for and about the therapist as it is about the patient.

When cognitive scientists use the term *narrative*, it is tied to a discussion of certain types of memory systems: how these systems have been isolated in tests and how their location in the brain was established.

Clinicians, on the other hand, are concerned with how to achieve successful outcomes in their treatments. They are concerned with how what they experience as success can be accounted for objectively. Broadly speaking, however, both neuroscientists and psychotherapists today seem to agree that narratives serve the need to remember, to create meaning, and to deal with future occurrences of similar events (Bruner, 1990; Schore, 1994; Squire, 1987). This is a good place to start, since the two disciplines concur.

As I will discuss later in this book, developmental research also suggests that core aspects of the person are organised in early parental experiences. These experiences allow the child to create a story of and about the self (Nelson, 1993). To fully understand how a therapist's memory works, we may have to know how these early stories about selfhood are formed and maintained, and in what way they are part of how we relate to the selfhood of others.

Ending my session with Brian

As my patient Brian is about to leave and we shake hands, I am still focused on his mood and his demeanour: what our last eye contact tells me, his parting words. Although the verbal dialoguing is over, we are still in session until he closes the door behind him.

All of this is more or less a repeat of what I have done hundreds and hundreds of times with patients, a ritual as well established in my memory as walking or driving a car, a procedure that requires no conscious initiation. But the ritual is also a chance of transitioning while Brian goes through his own ritual of departure, a chance to begin to take stock of what has transpired and what is different from when he entered my consulting room.

These observations may in fact prepare me for the next time I see Brian. For a few minutes, I can focus on what seems most relevant and what stands out about the session we just had. I may take notice of what he brought to my attention that I did not fully respond to. I may also notice how certain disclosures of his remind me of sessions when we first began the treatment.

Our session is still on my mind as I gradually begin to think of what is next on my schedule. And as I make my short notes, I am still relying on what has become so habitual that it easily could be called a technique. My processing of the session, however, remains intentional and focused on the one event in which I participated moments ago. Images and certain expressions, be they verbal or nonverbal, are vividly recalled even before I can translate them into words and theoretical formulations.

This phase of my work with Brian is exclusively internal and it serves the dual purpose of rehearsing and consolidating memory, on the one hand, while also restoring a sense of my personal boundaries, on the other; thoughts that belong to me as a person can now be reclaimed, something the intensely emotional involvement in sessions may not have permitted (Levine, 1994).

While my processing during the sessions with Brian seems to involve a heightened state of absorption and relational awareness, what happens postsession helps me review and consolidate information. It does so before many of the references to theory enter my mind. My previously held notions may then be revised and my attitudes modified if they impede my understanding of particular aspects of his psychology.

What we are dealing with, in other words, is a rather complex learning process. Simply in order to remember what occurs in a given therapy session, I must rely on already formed knowledge and beliefs that I had before I began working with Brian. But I must also retain information that is specific to my work with him. How well I do so will have an impact on how I continue our therapy and will be of critical importance to both of us.

CHAPTER TWO

The nature of subjectivity

When a group of psychotherapists in the 1970s began discussing subjectivity, neurocognitive research had yet to be an established field of study. Their explorations of what later became a theory of intersubjectivity were instead a reaction to the impersonal and procedural way that the relationship between therapist and patient had been approached by most psychotherapists, especially in psychodynamic circles. Following the advice in Freud's technique papers, therapists, especially during the postwar period when psychoanalysis was most popular, had sought to follow a basic rule of remaining physically abstinent and emotionally neutral when working with patients. The technique of avoiding emotional responses and not volunteering any opinions or personal disclosures was supposed to ensure an objective stance and eliminate the possibility of undue influence on the patient (Freud, 1912e).

For some time, however, it had been clear to many clinicians that remaining unresponsive had a negative impact on the treatment itself and that most patients interpreted such a stance as a lack of caring. Furthermore, there was a growing awareness that therapists' observations about the general functioning and psychopathology of their patients were far from objective; they were only one participant's version of a complex relational dynamic. The traditional view of the patient's transference as an inevitable but distorted perception of the therapist no longer seemed tenable (Lessem, 2005; Mitchell, 1988). The increasing consensus was that the therapist's perceptions of a patient's reality, accordingly, must themselves be subjective.

Based on such conclusions, many therapists, among them Robert Stolorow, Bernard Brandcraft, and George Atwood (1987), now argued that treatments are better understood as being based on subjective truth. They write:

> One source of interference with the quest for subjective truth is the presumption of an objective reality "known" by the therapist and "distorted" by the patient … . In general we find that therapists are most likely to invoke the concepts of objective reality and distortion when the patient's experiences contradict perceptions and beliefs that the therapist requires for his own well-being.

(p. 135)

To these therapists, subjective truth constitutes "a science of the *inter-subjective*" (Atwood & Stolorow, 1984, p. 41). Treatments are what they call "the interplay between the differently organised subjective worlds", in which the therapist functions as the observer to the patient as the observed. Together, the two participants form what Atwood and Stolorow call "an indissoluble psychological system" (pp. 41, 64).

This was a radical change—from the individual to the social, from the linear to the nonlinear—and it coincided with changes in many fields studying the mind (Bruner, Olver, & Greenfield, 1966). For the credibility of psychotherapy—and psychoanalysis, in particular—the new understanding also seemed to remove many of the scientific justifications for what happens in treatments and it deepened an already strained relationship with the research community (Winer, 1994). A longstanding medical legacy among U.S. practitioners trained analytically was questioned, as well as the established criteria for training. Limiting training in psychoanalysis to psychiatrists, thus making it an exclusive subspecialty of medicine, now seemed outdated and groundless (Schwartz, 1999).[1]

What proponents of intersubjective theory put in motion is now, in fact, at the cutting edge of research in cognitive neuroscience and important findings about memory. Current models describe the retrieval of certain forms of memory as an exclusively subjective experience. This memory system, identified as *episodic memory*, turns out to be based on how a person remembers the time, place, and sensory data of past experiences. Thus, episodic memory is how we retrieve first-person knowledge, which is fundamentally different from how we retrieve third-person knowledge of objective data.

These models for human memory emerged from research that is no longer hampered by the often-proclaimed division between actual and subjective experience in which the latter is regarded as beyond study (O'Connor, Moulin, & Cohen, 2008). Endel Tulving of the Rotman Research Institute in Canada, one of the pioneers in this type of research, articulated the importance of these developments in a conference paper on episodic memory in 2000 at the Royal Society in London (Tulving, 2002). He writes:

> Much of science begins as explorations of common sense, and much of science, if successful, ends if not in rejecting it then at least going far beyond it. The science of memory, although still in its formative years, is no exception to the general rule. Many findings yielded by research, and theoretical interpretations of them that were brought up at the Royal Society discussion meeting on episodic memory and appearing in print here, neatly illustrate this point. They transcend traditional thought.
>
> (p. 269)

Instead of identifying memory with information storage—to Tulving, a common-sense mistake when experimental data are missing—researchers are now exploring the complexities involved in retrieval of information once it has been encoded. This also means that subjectivity and statements referring to first-person experiences are no longer perceived as beyond scientific study. When retrieving a memory of a specific past event, the experience is an actual and observable event that can be measured and compared across subjects. Thus, there are now methods to quantify personal confidence in the experience of retrieving information: how much detail was being recollected and what was being recollected (Yonelinas, 2002).

For instance, when testing forty-eight subjects about their remembered responses, John Gardiner of the University of Sussex found that the subjects' descriptions "always included specific contextual details connected with the study-list presentation of test words" (2002, p. 13). This type of memory thus appeared to be quite specific and open to study.

How therapists remember

The therapist's role in treatments certainly relies on such first-person remembering of what occurred in previous sessions. For example, a fifty-nine-year-old male patient relates to the therapist about the care of his mother, for which he now is responsible. He had been in treatment for over five years and in this particular session he expresses satisfaction over the fact that she seems well taken care of. Due to his arrangements, she remains in the house where she raised him. As he reveals the costs of her care, the patient then wonders how long he will be able to keep this up financially, and he has a definite change in mood. He turns to the therapist, asking, "Do you think I am doing the right thing?" and the following dialogue ensues:

THERAPIST: Sounds to me that you are doing everything you can for her.

PATIENT: Yes, but soon I have used up all her money and now the money has to come from me.

THERAPIST: So what else would you want to do for her? (*Now remembering how the patient in an earlier phase of the treatment had vivid memories of the mother's ignoring his appeal for attention when he was a child. Among other recollections, he had remembered the time when he had spent days creating a model of the house in the country where they would visit her parents and the mother's ignoring his creation.*)

PATIENT: (*long pause, pushing back on his chair and abruptly changing the subject*) I always could relate better to men. At work, women have always complained that I ignored them and that I favour the men.

THERAPIST: Do you remember how you would tell me how alone you felt as a child, how your mother abandoned you? Maybe that has something to do with how you relate to women?

PATIENT: You are right. (*Seeming to remember particular scenes.*) From the time I was a teenager, I had these crushes on girls. I felt I was so unattractive. I thought I had no class. (*Pause.*) I remember trying to date this girl, Katelyn, in seventh grade. She wouldn't have anything to do with me. (*Pause.*)

THERAPIST: Like when your mother ignored you?

PATIENT: (*Clearly emotional.*) Look, my mother was so unhappy with everything! (*Pausing to take a deep breath.*) She seemed to hold herself above everyone else in the neighbourhood. She never felt we belonged. I must have been such a lonely kid.

After the initial discussion of the mother's care, the therapist here reminds the patient of what happened in an earlier stage of the treatment. This serves as a cue to adolescent memories and the kind of "mental time travel" that Tulving (2002) associates with episodic memory, with its vivid imagery and strong emotions. The patient's reliving of these past experiences was made possible by the therapist's remembering disclosures earlier in the treatment—disclosures that, for the therapist, were cued by the fact that the patient was now dealing with the relationship to his mother.

Such retrieval of past exchanges serves a critical role in treatments. And in order for this material to be solidly encoded, the therapist must have maintained a focused attention when the patient's experiences were introduced. Once encoded into the therapist's memory, however, the specifics of what the patient experienced can be retrieved at any stage in the treatment. There is less of a need to rely on a general interpretation; in this case, an understanding of the relationship to the mother. Theoretical formulations, such as *mother complex, castration anxiety*, and so on, would have identified the patient's disclosures with a label, but they would also have made the specific disclosures less accessible for retrieval at a later time. In the therapist's memory, a generic index, an interpretation, would have been tagged to the disclosures, thereby complicating the ability to cue the experience at a later time for the patient.

Fear of reductionism

Unfortunately, the recent findings about episodic memory appear to have escaped many psychotherapists, including those who were proposing the new intersubjective approaches. Once their theory was established, its originators resisted efforts to explore new data in the cognitive sciences (Stolorow, Brandchaft, & Atwood, 1987). On the contrary, as Robert Stolorow confirmed in a 2005 interview, they often viewed information about the brain as a form of reductionism, similar to Freud's, in which the therapy narrative tends to boil down to biological concepts (Doctors, 2005, p. 7).

When the same therapists describe the role of empathic attunement and the significance of the therapist's ability to cue the patient's experiences, they nevertheless touch upon important findings about subjectivity. Stolorow and colleagues write:

> To say that subjective reality is articulated, rather than discovered or created, not only acknowledges the contribution of the therapist's empathic attunement and interpretations in bringing these prereflective structures [the patient's "preformed meanings and organizing principles"] into awareness. It also takes into account the shaping of the reality by the therapist's organizing activity, because it is the therapist's psychological structures that delimit and circumscribe his capacity for specific empathic resonance.
>
> (Stolorow, Brandchaft, & Atwood, 1987, pp. 7–8)

What Stolorow and colleagues call "the patient's articulated subjective reality", is, in several aspects, a description of episodic memory. Furthermore, when they discuss how "prereflective structures" are brought into awareness, as described later in this chapter, they seem to describe what the researchers call *autonoetic awareness*, a conscious state specifically associated with episodic memory (Tulving & Lepage, 2000).[2]

Thus, by applying recent neuroscientific findings to these therapists' formulations, new levels of understanding are added to their observations—in particular, how the therapist's responses assist in accessing first-person knowledge. Although this understanding did not come from observations based on psychodynamic theory, it does in fact enhance a common therapeutic experience. Identifying observations made from a psychodynamic perspective in neurocognitive terms confirms rather than diminishes their relevance to the treatment process. It also makes it possible to more precisely formulate a theory for what in fact happens in a treatment.

The division that sees the sciences as "objective and factual" and psychoanalysis as "subjective and beyond empirical study" unfortunately continues to exist among therapists. Mark Solms and Oliver Turnbull, in a valiant attempt to reconcile current research and established clinical theory, still perpetuate the bias that subjectivity cannot be researched. In their book *The Brain and the Inner World* (2002), they write:

> Only matter—or only external perception—permits the reliability of *multiple observers*. This reduces the potential for *observer bias* that occurs with lone observers (like psychotherapists). Minds, by definition, cannot be observed by multiple observers. Similarly, from the external point of view we can study mental instruments of *nonhuman animals*. Nonhuman animals cannot provide verbal reports of their experiences.
>
> (p. 77; italics in original)

This is an outmoded distinction about research and assumes that only phenomena documented by several observers in independent settings can be studied. Their description of the problem leaves out findings based on anatomical and statistical evidence; it also ignores a now-viable option in researching first-person memory: the experience itself will manifest in measurable neurological changes, and its retrieval routes are open to study.[3] No longer do conventional distinctions hold true, in which recognition decisions (third person) are open to science while remember-know decisions (first person) are not. In both instances, objectivity can be achieved with equal accuracy (Gardiner, 2002). And if the findings from such research enhance the effectiveness of psychotherapy treatments, it must be possible to integrate these findings into a broader understanding of what happens in those treatments.

Subjectivity and psychoanalysis

Philosophically, subjectivity and what appears to be exclusively personal about certain experiences have been a conceptual challenge for thinkers and researchers for a long time. To some schools of thought, nothing can be said about the world other than that it exists because someone has an experience of it. Such subjectivism has often relied on hermeneutics rather than scientific explanations. Other philosophical schools insist that the external world exists on its own; it can only be described when relying on objective measures and is not influenced by the subject's state of mind. This school of thought therefore rejects any inference made by introspection (Bunge, 1999).

Under the influence of behaviourism, the second view prevailed in psychology well into the 1970s; human neuroanatomy was regarded as a merely passive organ responding to environmental contingencies. This view was overcome only with renewed explorations of the workings of the brain and the inner workings of the mind—again a credible field of research (Sternberg, 1996).

In psychoanalytic thinking, subjectivism has generally been viewed as something to avoid. Most of the supporting evidence for particular theories has traditionally come from sources in the natural sciences; these, in turn, would make analytic thinking and models free of subjective elements (Spence, 1982). The findings about episodic memory, therefore, represent a shift in which the old philosophical dilemma has to be reframed. Although autonoetic awareness is subjective, it is a significant aspect of explicit or conscious knowledge. The other conscious awareness, *semantic* or *noetic* knowing, can no longer be seen as exclusive to consciousness.

In theorising about the nature of psychoanalytic knowledge, therapists have certainly made subjectivist arguments, but mostly in reaction to arguments perceived to be too reductive. Assertions, for instance, that Freud's early formulations about human sexuality must be the sole basis for interpretations have been met with justified scepticism (Fisher & Greenberg, 1977).[4] In fact, many of the theoretical arguments and divisions into "schools" may be seen as theorists' failure to recognise the inevitable role of subjectivity in psychoanalytic formulations (Atwood & Stolorow, 1993).

In the writings of Heinz Kohut, the founder of self psychology, subjectivity is equated with what he terms "the scientific use of introspection and empathy", to him the only truly analytic way to access the patient's inner world (1959, p. 211).

> The inner world cannot be observed with the aid of our sensory organs. Our thoughts, wishes, feelings, and fantasies cannot be smelled, heard, or touched. They have no existence in physical space, and yet they are real, and we can observe them as they occur

in time: through introspection in ourselves, and through empathy
(i.e., vicarious introspection) in others.

(pp. 205–206)

By claiming psychoanalysis as an independent discipline and a science
in its own right, Kohut opposed earlier reliance on biologically based
theories. Instead, he insisted that data are about the patient's inner world
and can only be accessed empathically, via the therapist's introspection
extended vicariously to the inner world of the patient. Empathy, as he
defines it, thus renewed the interest in certain clinical responses that
therapists have described for some time. This also meant that certain
interpretations based on Freud's drive theory fell out of fashion.

Kohut's formulations come close to how autonoetic awareness is
described in today's neuroscientific literature (Tulving, 2002). Empathy
involves withdrawal from the immediate focus on external events so
that the therapist can be immersed in the patient's private experiences.[5]
Although Kohut's formulations do not account for the particular memory systems involved in such immersions—most of his formulations
predate the research on episodic memory—they can now be tied to the
particular memory system in patients that therapists, in their empathic
responses, cue.

Dependence on research

The credibility of all psychotherapy formulations depends on research
in disciplines such as neuroscience. There may still be resistance to rec-
ognising this fact; however, such dependence has been evident since
the inception of psychoanalysis. Supporting findings in life sciences,
such as biology and medicine, allowed for many of the critical early
discoveries, such as the primacy of early experiences and the power
of fixation and regression (Sulloway, 1979). They also, unfortunately,
bolstered the conviction that certain of Freud's formulations—such as
the dream theory, Oedipal strivings, and the origin of homosexuality—
were "proven", even when contradicted by new data.

Over time, many of these claims of proof have been dismissed. In a
seminal review published in 1977, Seymour Fisher and Roger Greenberg,
two researchers from the Upstate Medical Center of the State University
of New York, examined such formulations. They also took a critical look

at the justifications for Freud's formulations about therapeutic actions and found that they all lacked substantial support in research. In terms of his dream theory, for example, they conclude:

> First, there is no rationale for approaching a dream as if it were a container for a secret wish buried under layers of concealment. It is closer to the facts to regard it as a fantasy produced in a special state of consciousness and containing a spectrum of information about the dreamer.

> (p. 68)

Later reviews, from 1988 and 1993 by Adolf Grünbaum, a philosophy of science professor at the University of Pittsburgh, arrived as similar results, with less forgiving conclusions. Grünbaum writes:

> [I]f psychoanalysis is to have a future as a scientific enterprise, it very probably does not lie with the clinical case history method, but with other testing designs … . Yet to this day the case study method has been the source of evidence adduced by the vast majority of therapists who claim support for their theory of psychopathology, dream theory, theory of slips, and theory of psychosexual development.

> (1993, p. 162)

Today we are left with the sobering realisation that many of Freud's theories—as well as other, less orthodox, formulations—have been proven wrong, and that others are impossible to verify (Bornstein, 2001). And even if some hypotheses could be tested in controlled studies using data from nonanalytic settings, as proposed by Grünbaum, a more feasible approach would be to explore how to integrate current scientific findings applicable to the practice of psychotherapy (Flanagan, 1984; Grünbaum, 1993). Not only would such an approach reestablish the credibility of how treatments are approached, it would also anchor formulations about them in a framework of clearly defined concepts.

The explicit memory systems

In approaching the data about episodic memory in treatments, many of the older, established notions about the general function of memory need to be revised. Even focusing solely on long-term conscious memory, at least two such explicit memory systems have been identified within the research community, aside from several implicit or unconscious systems described in the previous chapter.

As Daniel Schacter, Anthony Wagner, and Randy Buckner (2000) define them, several distinct memory systems exist, each allowing us to perform a variety of tasks or functions within a particular class or domain. Thus, different kinds of information are handled within their own domains. One way these classes or domains are determined is through what they call *class inclusion*: when changes in brain states impair performance within a particular system, but have little or no effect on the performance outside its domain, that is evidence of a selective operation of that particular system (p. 628).

The first explicit memory system is called *semantic* and mainly deals with factual, third-person information. It appears to develop during the second year of life and is the basis for what we generally think of as factual knowledge, including facts, vocabulary, concepts, and grammar, and what in everyday life is considered as objective and rational (p. 632).

Episodic memory, on the other hand, is directly related to first-person knowledge and appears to develop first in the preschool years. It is associated with the hippocampal region and, in particular, areas of the prefrontal cortex (LeDoux, 2002). These prefrontal cortical functions continue to mature through the preadolescent period, resulting in increased connectivity between temporal lobe structures and the prefrontal cortex (Nelson, de Haan, & Thomas, 2006). In their mature form, these functions are related to self-awareness and the ability to engage in introspection regarding one's own thoughts. Through such self-awareness, the relation between self and its social environment is therefore maintained (Stuss & Benson, 1986).

The prefrontal cortical functions are also, through more posterior connections, related to language, attention, and motor control (Wheeler, Stuss, & Tulving, 1997). To some researchers, episodic memory is, therefore, an exclusively human form of consciousness (Ramachandran, 2011; Siegel, 2003).[6] However, this conclusion is contradicted by research on food-storing birds and rats (Clayton & Dickinson, 1998; Wood, Dudchenko & Eichenbaum, 1999). Although undetectable in many animal species, episodic memory may manifest also in nonlinguistic behaviour.

Proposed by Endel Tulving, the explicit memory systems are based on two kinds of consciousness: *noetic*, or knowing, and *autonoetic*, or self-knowing (1983). Episodic memory, as related to autonoetic consciousness, is called on when "a child remembers what happened at a friend's birthday party the day before, a young lover remembers the expression of his beloved's face in the moonlight, the scientist remembers the first time when a speaker at a conference mentioned her work, and so on"(Tulving & Lapage, 2000, p. 211). As "mental time travel", episodic memory provides a sense of being able to recollect one's self at a particular time in the past, and such recall is accompanied by a strong, subjective sense of certainty (Wheeler, Stuss, & Tulving, 1997).

Initial discoveries

The initial discoveries about memory's being composed of several systems were made as the result of studies of patients with brain lesions (Schacter, 1996). In performing experimental surgery for epileptic seizures on patient HM in 1953, his doctor, William Beecher Scoville, removed critical deep brain structures on both sides of the medial temporal lobes. Although the operation appeared to cure the seizures, and, in fact, increase the patient's intelligence, the result of the surgery was a profound amnesia. HM was unable to recognise the hospital staff and he often forgot about the most recent meal he had eaten (Scoville & Milner, 1957).

In HM's surgery, most of the hippocampus, the amygdala, and other areas of temporal cortex were removed. The study of other patients, with less extensive damage to the hippocampus, confirmed its role for remembering personal experiences (Janowsky, Shimamura, & Squire, 1989). Thus, hippocampal functions appeared critical to remembering events in the recent past and to encoding new episodic memory, while semantic memory for facts seemed to function independently of source encoding.

Neuropsychological testing and animal studies later confirmed that damage to the hippocampal areas was associated with loss of episodic memory, although semantic memory functions remained intact (Wheeler, 2000). In fact, source amnesia appeared to be a characteristic of semantic memory and to distinguish it from episodic memory (Schacter, 1996).

Other studies confirmed memory loss in one, but not the other, explicit memory system. Focusing on semantic memory, McKenna and Warrington established that there is "no dispute about the multiplicity of semantic *access* routes, each of great complexity, as witnessed by the different stages of analyses from simple sensory stages to final recognition of semantic properties" (2000, p. 374; emphasis in original). Similarly, in studies by Patterson and Hodges, damage to specific brain regions "will disrupt semantic memory without harming other cognitive functions directly", thus isolating semantic memory as a distinct system (2000, p. 313).

In recent studies using neuroimaging techniques with healthy subjects, further insights about how the memory systems interact have surfaced. Using one such technique, event-related fMRI, a Harvard team lead by Anthony Wagner and Daniel Schacter established that regions within the prefrontal cortex play an important role in episodic encoding. They also found medial temporal lobe activation with episodic encoding (Wagner et al., 1998). Based on the new findings, Schacter, Wagner, and Buckner conclude that "a number of the same regions are involved in multiple systems" (2000, p. 638). However, the left hemisphere appears more involved in the encoding of episodic memory, while retrieval relies in the right frontal lobes (Mayes & Roberts, 2001). Episodic memory traces also appear to be visual and emotional (ibid.).

The two conscious or explicit memory systems are, therefore, primarily located in the prefrontal cortex and depend on communication and integration via the corpus callosum, which connects the two halves of the brain (Cozolino, 2006).

Factual memory defined

Today, semantic memory is generally understood as a network of associations and concepts on which we base our knowledge of the world: word meanings, categories, facts, and propositions. Everything from "a bird has wings and two legs" to the exact address where we live are stored semantically (Schacter, 1996). So are various rules and conventions, such as what to avoid and what produces pleasure (Solms & Turnbull, 2002).

One of the striking characteristics of semantic memory is *source amnesia*, referring to the fact that semantic memory does not preserve knowledge of its encoding contexts. Many of us know that Paris is the capital of France, but it is usually impossible for a person to recollect when he or she originally learned this fact. Semantic knowledge presents us with the meanings of words, numbers, and facts, but without our remembering how we learned about them (Anderson & Conway, 1997). Semantic knowledge is, essentially, what we remember in third-person statements; it has no references to self and personal experience. It may in fact be "acquired through repeated exposure to the same information in a variety of different contexts, such that memory for the information becomes aspecific" (Rugg, Johnson, Park, & Uncapher, 2008, p. 339).

Temporal memory defined

Retrieval from episodic memory, on the other hand, depends on source information from the initial encoding (Wheeler, 2000). Episodic memory is the only form of memory oriented toward the past; and the cues for retrieval are temporal or spatial—or both—in nature (Tulving, 2000). If, for instance, we wish to remember what we had for breakfast yesterday morning, the most probable way to retrieve this information is to recall what we did before having it or where we were at that time.

The unique characteristics of episodic memory are the reliving of scenes from the past and the fact that what is remembered can only have been experienced by the person whose memory it is. The personal nature of what is relived is what makes it possible to reflect on these events in the present, with a sense of having been a witness to them (Mayes & Roberts, 2001). Episodic memory may, in fact, give us tools to reexamine convictions we have about the world and ourselves because they are based on our own experiences. Christoph Hoerl, a philosopher at the University of Warwick, explains:

> The thought here is that episodic memory has a specific role to play in the kinds of knowledge about events and event sequences a person can be said to possess. In particular, a person's ability to remember particular past events can explain a sense in which she can grasp why it is right for her to hold certain generalized beliefs. A person who remembers how things went on previous occasions knows not just that there are instances of which certain generalizations are true. Her episodic memory will also provide her with a grasp of how she is right about those instances, that is, how she is in a position to know about them.

> (2001, p. 323)

Episodic memory is, therefore, directly linked to knowledge of our personal history and how this history, although unique and subjective, is the basis for the values and beliefs we rely on. It is a way to review these convictions, by either confirming them or having to adjust them. And this capacity allows for projections into the future by reflecting "what one's own experiences might be like at a later time" (Wheeler, Stuss, & Tulving, 1997, p. 335).

Since episodic memory appears to be based on establishing associations between different elements of an event, several studies indicate that the memory traces tend to weaken over time (Schacter, 1996). To Martin Conway of the University of Durham, England, most episodic traces are lost within a twenty-four-hour period and are only retained when linked to more permanent autobiographical knowledge (2001).[7] Daniel Siegel comes to a similar conclusion and places a great deal of emphasis on what he calls "cortical integration" via autobiographical narratives as a means of making episodic memory traces permanent (2003).

Endel Tulving, for his part, holds out the possibility that certain episodic information endures long term, but this would require a certain state of consciousness. Only when "the rememberer's belief that the memory is a more or less true replica of the original event, even if only a fragmented and hazy one, as well as the belief that the event is part of his own past" would such enduring storage be possible (1983, p. 127).

Real-life memory

In our everyday lives, one memory system will often be used in conjunction with or subsequent to others. Generally, research projects do not test situations where such dual use occurs. In this sense, clinical practice is much closer to everyday life, where memory functions are not isolated by a test protocol. Thus, the specific interplay between memory systems often goes unnoticed by researchers, or they regard it as secondary. Conversely, if clinical experience were the sole source for our information, a unitary memory system would still appear to be the most logical assumption.

The questions of how semantic and episodic structures overlap and in what order they may be activated have so far found several possible answers. Tulving (2002), in an attempt to resolve the issue of sequencing, argues for a third explicit memory system, *perceptual memory*, which is mainly concerned with the form and structure of words and objects. According to this model, once encoded, some information may remain exclusively perceptual and other information may progress into the semantic system; but for information to reach the episodic system it first has to be stored in the previous two (ibid.). In this model, called the SPI model, the relationships among the systems are what he calls "process-specific": encoding occurs serially (S), storage is parallel (P), and retrieval is independent (I) (p. 275). In other words, encoding of information would occur in a linear sequence, one operation at a time, while storage happens with multiple operations going on simultaneously.

Further arguing for this model, Tulving and Markowitsch (1998) point to the fact that in a group of young adults suffering from anterograde amnesia due to damage to the bilateral hippocampus, episodic memory appeared affected but semantic memory remained intact. They conclude that episodic memory must be "a unique extension of semantic memory" and that "the hippocampus is necessary for remembering past experiences and the remaining medial temporal lobe (MTL) regions are necessary for the learning of factual information" (p. 199).

In adding a perceptual system, Tulving and colleagues thus maintain the idea that true episodic recall assumes that the information was first processed semantically. Other researchers, such as Larry Squire and S. M. Zola (1996) of the University of California–San Diego, are more inclined to view the two memory systems as working in parallel.

Retrieval and its mechanisms

Of particular importance to our understanding of conscious memories is research on the mechanism responsible for retrieval of these memories and the specific interaction between encoding and retrieval. The *encoding specificity principle*, first formulated by Tulving and his colleague D. M. Thomson in the 1970s, establishes that a retrieval cue can only be effective when the information in the cue was incorporated in the trace of the target event (1973). In other words, the particular way a person thinks about an event determines the likelihood of later recalling it: the cue must reinstate or match the original encoding. Daniel Schacter describes *cueing* as follows:

> These considerations [the nature of the initial encoding] suggest that the way we perceive and think about an event plays a major

role in determining what cues will later elicit recollection of the experience. But it is not the literal similarity between encoding and retrieval conditions that is the crucial determinant of the explicit memory. Rather, what matters most is whether a retrieval cue reinstates a person's subjective perception of an event, including whatever thoughts, fantasies or inferences occurred at the time of encoding.

(1996, p. 61)

Remembering, therefore, is what happens when some of the previously encoded data are activated by the associated cue. Typically, such retrieval involves communication between the left and the right sides of the brain. However, retrieval is not a simple replication of the stored information. It is a new construct that is only *similar enough* to the neural pattern previously encoded; thus, the potential for omissions, misplacements, and distortions (Schacter, 1996).

Researchers have also described different kinds of cues and forms of retrieval. *Contextual cues* rely on external context, such as place, time, and other distinctive conditions (Brown & Craik, 2000). Also called *environmental supports*, they comprise one of the major ways to cue episodic memories—for instance, two British researchers, D. R. Godden and A. D. Baddeley, showed that scuba divers had better recall of what they had learned when diving when they were again on location, under water (1975).

State-dependent cues, on the other hand, depend on a person's mental state at the time of encoding: thus, retrieval will occur when the original mental state is reexperienced (Brown & Craik, 2000). Even if the cue is weak, the marked presence of mood or arousal during encoding will cue similar thoughts and feelings. When patients enter a depression, for example, chances are that they will remember the dysphoric thoughts associated with being depressed in the past and be prone to repeating these thoughts as a well-rehearsed story. (See Chapter Four for discussion of such stories.)

Depending on the rememberer's conscious participation and the effort involved, the retrieval process can be described in several ways. *Associative retrieval* is often involuntary; all kinds of experiences spring to mind in response to simple environmental cues. The reliability of this type of retrieval is usually very low and elements of fantasy are common. *Strategic retrieval*, on the other hand, is less dependent on environment and is consciously guided. Often laborious, it consists of a search for information, such as the name of a person or a place visited. As detailed in the next chapter, this type of retrieval seems particularly related to autobiographical memory (Conway, 2004).

How children remember

Another contributing area of research is memory in small children. The findings about episodic memory initially caused difficulties for researchers studying early stages of life who have found it missing in young children. Most of the information retained from experience is, for the first years of life, absorbed into what Katherine Nelson of CUNY Graduate Center call *generic event memory* (Hudson & Nelson, 1986). (See Chapter Four for further discussion.)

But even if their memories do not persist into adulthood, children seem to remember novel events in detail. Thus, their memories appear different from semantic or episodic memories and researchers specialising in small children have demonstrated that early memories are formed by repeated exposure to similar events and a sense of "things happen this way" rather than the episodic sense of "something happened that *one* time" (p. 7; emphasis in original). To Michael Rugg, Jeffrey Johnson, Heekyeong Park, and Melina Uncapher of the Center for Neurobiology of Learning and Memory at the University of California–Irving (2008), this type of memory is based on a sense of familiarity and "can support simple recognition, recency, or frequency judgement, but provides no access to qualitative information about prior event, and hence no specific information about where or when the event occurred" (p. 339).

At what point, then, are children able to form episodic memories? Katherine Nelson and Robyn Fivush find this to be a still-open question. They write:

> Although we now know a great deal about memory in infants and early childhood, some puzzles of early memory remain. In particular, it is difficult to establish whether children remember a specific episode as an episodic memory in Tulving's sense—that is, as having happened at a particular place and in a particular time.

> (2000, p. 285)

The important point here is that the young child's limited memory system must be economical. Only what is significant, by occurring repeatedly, will be stored consciously. In an earlier paper, Nelsen seems to answer this question by pointing to an entirely different system:

> The solution for a limited memory system is either to integrate the new information as part of the generic system or to keep the novel memory in a separate, temporary, episodic memory for a given amount of time to determine if it is the first of a series of recurrent events and thus should become part of the generic system. Then, if the event reoccurs, the memory may be transferred to the more permanent generic memory system. If a similar event does not occur during the test period, the episode is dropped from memory as of no adaptive significance.
>
> (1993, p. 11)

As I discuss in more detail when dealing with narrative memory (Chapter Four), the memory functions that Nelson finds in her research on children have much in common with *scripts*, simple narrative structures first described by Schank and Abelson in 1975 and later described as "a list of events that compose a stereotypical episode" (Schank, 1999, p. 4). Such scripts, to Nelson, provide "a schema derived from experience that sketches a general outline of a familiar event without providing details of the specific time or place when such an event happened, whether once or many times" (1993, p. 7). A script, then, specifies the sequence of actions, while leaving what Nelson calls "empty slots for roles and props that may be filled in with default values, in the absence of specifications" (ibid.). Scripts are the child's way of generalising from past experience, often told in the timeless present tense using the second-person pronoun. Scripts do not report what its owner did but "what one does" in a certain type of situation (Hoerl, 2001, p. 321).

Scripts, it turns out, are also ways of remembering that we, as adults, rely on. They are important ingredients in how therapists apply certain generalised knowledge to what a patient may be speaking about. Thus scripts must, in themselves, be a type of memory that develops earlier than most other types and that takes on particular narrative functions in adults.

Current status of the research

The lack of truly episodic memory in small children thus points to its having a rather late arrival developmentally. Daniel Siegel (2003) of the UCLA School of Medicine offers one explanation for this. Results from recent brain imaging studies point to a slow maturation of the orbitofrontal cortex, first beginning during the preschool years. The delayed developments in this region would thus explain how experience "continues to shape the way we come to understand ourselves and the world in which we live throughout the lifespan" (p. 26). Siegel bases this conclusion on "a number of highly relevant processes subsumed by these coordinating areas of the brain that are relevant to autobiographical memory" (ibid.). Located in the prefrontal cortex, the orbitofrontal cortex "sits in the junction of the other limbic structures (including the anterior cingulated cortex, hippocampus, and amygdala), the associational regions of the neocortex, and the brain stem" (ibid.).

The slow maturation of these areas would explain Nelson's conclusion that the acquisition of autobiographical memory—typically, in a transitional two- to five-year-old age range—represents a more specific, long-lasting, and particular achievement fostered by social conditions (2000). The child's representational system is now accessible to linguistic formulations presented by its caregivers. In turn, the child achieves a sense of having a unique and personal history.

This also suggests that the capacity to form narratives is an essential part of autobiographical memory. This capacity is what makes children able to form lasting explicit memories. Tied to the development of language, autobiographical memory thus serves a dual function: while providing mental representations of the self, it serves a need to communicate. However, an equally important element in forming autobiographical narratives is the capacity to encode and retrieve episodic memories. The two are intimately linked, to the point that some researchers, among them Daniel Siegel (2003), regard them as the same.

One reason for the confusing designation of *episodic* versus *autobiographical* memory may be related to the visual and subjective aspects of episodic memory. In terms of retrieval, episodic memory is accessed via temporal and spatial cues and is highly context dependent. Autobiographical memory requires a strategic and effortful retrieval, often consciously guided (see Chapter Three for a fuller discussion). It also includes some type of narrative component, and thus appears related to language (Conway, 2004).[8]

Less confusion exists over semantic memory and episodic memory as two separate systems. As we have seen, the research no longer supports a unitary understanding of what we remember consciously (Schacter, Wagner, & Buckner, 2000). This fact has clear implications for understanding what occurs in psychotherapy treatments. As the research into first-person knowledge shows, autonoetic awareness, on which the therapist as well as the patient relies, is no longer beyond empirical study. Findings about episodic memory—and the particulars of retrieval—shed new light on previously mislabelled aspects of what happens in psychotherapy. Autonoetic awareness and mental time travel, as the hallmarks of episodic memory, are prerequisites in treatments.

Many factors may be involved in how well such episodic traces are maintained in long-term memory. As discussed in the previous chapter, encoding may be aided by prolonged and heightened attention, a process called *consolidation*. Memory traces can then resist decay for a long time, especially those traces that can be claimed to belong to one's own past experiences (Tulving, 1983; Wheeler, 2000). And in terms of retrieval, current research appears to confirm that memory about personal experiences in the past—at least in humans—is associated with autonoetic awareness and the ability to withdraw attention from the immediate sensory environment. Via appropriate cues, the sensual and visual information in those traces are then brought into consciousness.

As new data has surfaced, it has nevertheless become clear that semantic and episodic memory have certain features in common. Both represent large and complex systems, with seemingly unlimited capacity. When stored in both, information is accessible via a wide range of queries, prompts, and cues (Wheeler, Stuss, & Tulving, 1997). In some instances, semantic memory may also handle facts that directly involve the rememberer, even if, as the Tulving group puts it, "some of this system's operations are sluggish in the absence of support from episodic memory" (p. 333):

> Recollection of episodic information, by contrast, is not merely an objective account of what is, what has happened, or what one has seen, heard, or thought. It involves remembering by re-experiencing and mentally traveling back in time. Its essence lies in the subjective

feeling that, in the present experience, one is re-experiencing something that has happened before in one's life. It is rooted in autonoetic awareness and in the belief that the self doing the experiencing now is the same self that did it originally.

(p. 349)

The neural underpinning for the two explicit memory systems is also most likely similar. The medial temporal lobes, prefrontal cortex, and parts of the parietal cortex appear to be involved in retrieval of episodic memory (Watrous, Tandon, Connor, Pieters, & Ekstrom, 2013). For semantic memory, on the other hand, the hippocampal areas are less implicated, while the parahippocampal gyrus appears to be more directly involved (Yonelinas, 2002).

Clinical translation of the research

When applied to the clinical setting, remembering semantically and having noetic knowledge means that the therapist will remember the patient's name, his or her age, marital status, place of birth, and occupation, as well as facts related to abstract knowledge, such as general health, psychological symptoms, and a possible diagnosis. In fact, case reports include much of this type of factual information that the therapist assembles as part of assessments and in progress notes.

Semantic memory allows the therapist to hypothesise about causes and to conceptualise from internal and external data about the patient. At the same time, the therapist will most likely try to determine if the overall picture presented by the patient's disclosures is true or false— all part of the therapist's semantic knowledge. When applied to what a therapist does, such semantic knowledge represents a cognitive stance that is manifest in the ways the therapist explains, informs, and educates.

Episodic memory, on the other hand, is at the core of the treatment process. When the patient accesses troublesome or traumatic experiences from the past, he or she relies on episodic recall articulated as first-person knowledge. And when the therapist remembers what occurred in a given session, it is because he or she retrieved episodic memory material. The case of the fifty-nine-year-old male patient remembering his painful crushes as a teenager, noted earlier in this chapter, is based on *autonoetic awareness*, or self-knowing. Cued by the therapist, who recalled what was revealed in an earlier phase of the analysis, the patient's memories are articulated in the type of first-person statements that are particular to episodic recall.

Autonoetic awareness, the basis for episodic memories, thus makes it possible via cues to remember personally experienced events as placed in a distinct time and place. The resulting episodic memories are the basis for much of what is remembered by the therapist from a given session. For the patient, being able to travel mentally in time allows for a fuller narrative about the self to emerge from the encounter.

As discussed in detail in Part II, what I call *session-related memory* gives therapists only a partial account of what transpired in a treatment, even when consisting of well-preserved episodic traces. It represents what they were able to retrieve explicitly about a treatment and will only account for a small fraction of what transpired.[9] For the patient, session-related memory is intimately linked to his or her developing self-narrative and the integration of autobiographical knowledge. Thus, session-related memory is bound to have a different purpose and to be stored in a different context for each person involved in the therapy.

The traces of session-related memory are also highly dependent on contextual cues for their retrieval and may not be accessible apart from such cues. For the therapist, the initial encoding depends on the strength of his or her attention and his or her rapport with the patient, a fact that probably accounts for much of the variation in how therapists describe their treatments. Optimal emotional arousal assists in the encoding of the therapist's memory and consolidates otherwise lost traces (McGaugh, 1995).

Similarly, patients' memories of what occurs from session to session will depend primarily on contextual cues. In some instances, mood may also trigger their memories. According to Schacter, such *mood-congruent retrieval* is the result of state-dependent cues; he demonstrates that "sad moods make it easier to remember negative experiences, like failure and rejection, whereas happy moods make it easier to remember pleasant experiences, like success and acceptance" (1996, p. 211).

Thus depression, when unrecognised and unattended, may damage the patient's rapport with the therapist and become a major obstacle in the treatment. As shown in Chapter Four, such mood-congruent recall brings about a psychological regression in which old and well-rehearsed stories are disclosed. This is often an important aspect of the initial phase of a treatment. The therapist's own mood-congruent retrieval has the same potential of activating previously encoded material; thus, it is often associated with countertransference. As such, it may also be induced by the patient's mood—yet another way for the therapist to access information (Harris, 2005; Safran & Muran, 2000).

Episodic memory and psychotherapy

For the patient, autonoetic awareness and access to episodic memory is not only a necessary function for participating in a treatment; it is also what allows him or her to reflect on past experiences and imagine what they may mean for the future. Being in treatment requires being able to travel mentally in time, and, as the case vignette in the beginning of this chapter shows, autonoetic consciousness is what brings past experiences into a new focus. It is a source upon which lasting psychological change will occur and must tie what Conway calls "experience-near, sensory-perceptual knowledge" to the goals, explicit or implicit, of the therapy itself (2002, pp. 53–54).

For the therapist, episodic memory is the foundation for what he or she remembers from session to session. The consistency of the clinical setting, and the fact that therapist and patient meet with great regularity and for the same amount of time, is one of the main reasons for effective cueing of episodic memory. Through this type of memory in particular, past experiences are frequently open to retrieval with a high degree of accuracy. For the therapist, note taking, even when more semantic in nature, may further facilitate the encoding process and serve to cue new information.

For reporting on entire cases, however, the information at the therapist's disposal, now out of context, will be incomplete by necessity (see discussion in Chapter Eight). And as I will discuss in Chapter Four, much of the visual and sensual information that is the foundation for episodic memory will be contained in a narrative that often precludes the therapist's own communications and responses.

This also means that we have amazingly little knowledge about the particular kinds of cues therapists use to access their own memory traces and their information about each treatment in which they participate. Especially when it comes to what I have called *case-related memory* and how much a therapist is capable of remembering about an entire treatment, there is no actual data from research.

For the purpose of training future therapists, here is a worthy project for research: How does the individual therapist access what happens over the duration of a treatment? Contextual cues to episodic memory are no doubt implicated, but can empathic responsiveness and attunement, as a form of *state-dependent cueing*, also provide critical ways to access patient information more long term?

CHAPTER THREE

Retrieving history of the self

If autonoetic awareness is the necessary state of mind for episodic memories to form, then, by definition, access to these memories also involves the self as an agent and as a recipient. Episodic memory is in fact uniquely concerned with what I, the rememberer, believe to have occurred in my presence at a particular time and place. The retrieval, when it involves extensive personal experiences, is therefore autobiographical knowledge and, depending on the person's capacity to maintain and continually update a narrative about this knowledge, will make available a history of his or her sense of self (Siegel, 2003).

The need for consistency, for self-cohesion, makes this retrieval particularly demanding, especially when experiences from a distant past are involved; the concern is no longer remembering a dinner date, the name of a place visited, or a concert attended. To some researchers, this means that remembering our own autobiography ought to be treated as different from episodic memory, or at least as a distinct extension of it. It appears to demand more effort and more complex cueing than retrieving some basic facts from what has been stored semantically (Gazzaniga, 2008). And for the retrieval to result in a sense of expanded self-knowledge, the rememberer at some point would have to confront the paradox that the current self may, in fact, be involved in different life circumstances and have different goals and aspirations than the self remembered (Conway, 2001). Some type of reconstruction is demanded; a unified knowledge structure has to emerge and, along with it, a narrative about a uniquely experienced history (Kihlstrom & Klein, 1997).

But reconstruction involving the sense of self also assumes a level of verbal ability absent during the early years of life (Nelson, 1993). The acquisition of episodic and autobiographical memory develops slowly, while simple scripts of general events and semantic memory predate it. William Friedman of Oberlin College:

> With some help, even 4-year-olds can reconstruct past times on a time-scale basis they understand, parts of the day. But not until nearly 10 years of age can children consistently use their knowledge of the calendar to compare locations in the past. In contrast, children have access to distance-based processes at least by 4 or 5 years of age, and they can use this information to judge relative distances in the past with some accuracy. These abilities show that information about ages of memories is present early in development.
>
> (2001, pp. 162–163)

Whether it is the child's knowledge of the calendar that allows for autobiographical narratives to be the basis for self-knowledge or the tutorial relationship with a parent caregiver, full participation of certain brain structures is required (Nelson & Fivush, 2000). As Daniel Siegel (2010b) points out, encoding of explicit memories involves the hippocampus, an area of the brain located in the medial temporal lobe that grows throughout the life cycle. However, compared to other brain structures, this growth appears to happen slowly. In order for a person to form full autobiographical narratives, a certain accumulation of explicit memory may be required and the full participation of hippocampal functions.[1]

Autobiographical memories also seem to go through changes as they age. Since they are usually expressed in some type of narrative, the perspective may eventually change from a first-person (or field) point of view to a third-person (or observer) point of view (Rice & Rubin, 2011). Experiences with a strong emotional impact are usually encoded in a first-person perspective; they are prereflective rather than reflective (Tagini & Raffone, 2009). However, in older adults memories of the past may become *semanticised*; that is, what is being remembered are semantic details rather than distinct temporal or spacial contexts (Levine,

Svoboda, Hay, Winocur, & Moscovitch, 2002). This phenomenon, discussed in more detail in Chapter Four, appears closely tied to changes in the narrative component of the person's autobiography.

Whatever designation we give to retrieving our history and maintaining a sense of self, this type of memory has a direct bearing on the practice of psychotherapy and formulations about it. However, the idea asserted by the psychoanalytic pioneers, that the patient's past can be replicated in the treatment, can no longer be maintained without major revisions. Any reconstruction of the past can only be the result of what Roy Schafer of Columbia University's Center for Psychoanalytic Training and Research calls "remembering differently" (1983, p. 194). He writes:

> Because the reconstruction of the psychoanalytic past necessarily takes place in the here and now clinical dialogue, it remains an interpretable and reinterpretable feature of that here and now. This means that the past is always taken as that which is currently being told in one or another conflictual analytic context. Ideally, in each analytic undertaking, one continues to interpret the biases and limits not only of histories initially presented but of those that have been developed previously in that very analysis and have now come into question.
>
> (ibid.)

The therapeutic reconstruction is, therefore, never complete. At best, it will consist of vividly experienced but highly selective memories that must be incorporated into an authentic narrative about the treatment in order to be retained. For the patient, autobiographical memory is, then, about the self as experienced in the therapy and through it. Although fluid and paradoxical at times, at least initially, these experiences are usually prereflective, often spontaneously emerging memories, and experienced in a first-person—i.e., field—point of view. For the therapist, on the other hand, the reconstruction will be remembered in an observer perspective as a series of relational events encoded episodically.

An initial session

The following case vignette illustrates how autobiographical information surfaces in a therapeutic treatment and is an example of an intensely experienced recollection accessed even in the initial interview. The therapist is trying to establish how the patient, a thirty-seven-year-old man struggling with depression, experienced certain family members. The therapist asks, "What was your mother like when you were a child?" The patient mentions that his mother went to work when he was seven. Before this, she stayed home with him and his brother, who was three years older. After a short pause, the patient resumes:

PATIENT: Actually, my mother was quite abusive to me when I was growing up.

THERAPIST: Could you tell me what you remember about it?

PATIENT: She would often scold me for being slow. I remember one time I was eating my breakfast in the kitchen where she always served it, on a little fold-out table near the door. She suddenly came up behind me and kicked the chair I was sitting on so it fell over. I fell backwards and must have banged my head on the floor. I can't remember what happened next but I know it really hurt. But she just continued to call me names, told me what a slowpoke I was. It was terrible.

THERAPIST: You must have been in shock that she did this. How old do you think you were?

PATIENT: I don't know. (*Now clearly upset.*) Perhaps five or six. My brother must have already left for school. (*Pause. Tries to regain his composure, turns to the therapist, still emotional.*) How could she do this? I hadn't done anything to her. She knew I couldn't see her. How could she do this? I hadn't done *anything*.

This type of retrieval will not occur in every therapy session and is certainly unusual in an initial session. But reliving a traumatic memory from childhood, as this patient does, is in general a remarkable event and often follows a particular test protocol described by researchers studying autobiographical memory.

In this case, the therapist stumbles upon a specific retrieval cue when asking for a description of the patient's mother. For the patient, the cueing activates what researchers who developed this protocol call a *recollective experience*. He vividly remembers a particular event in childhood that had a strong and lasting emotional impact.

Cueing and autobiography

The protocol was developed by Martin Conway and Airkaterina Fthenaki (2000), two English researchers specialising in the study of autobiographical memory. One of the subjects of their research articulated this typical sequence when the cue "cinema" was used:

> When did I go to the cinema a lot? When I was a student. I lived in a student hall near Russell Square and we used to go to the art cinema there and I remember sitting in the dark in a big red seat watching *The Spirit of the Beehive*.

> (p. 305)

The Conway and Fthenaki protocol has five components: (1) elaboration of the cue ("When did I go to the cinema a lot?"); (2) access to a particular period of life ("When I was a student"); (3) information common to the lifetime period ("I lived in a student hall near Russell Square"); (4) access to a general event ("we used to go to the art cinema"); and (5) access to sensory-perceptual details ("sitting in a big red seat") and other event-specific knowledge ("watching *The Spirit of the Beehive*").

What the protocol shows is commonplace in therapeutic treatment but often arrived at intuitively. A patient's autobiographical retrieval, requiring mutual effort on the part of therapist and patient, allows access to heavily emotion-laden memories and often facilitates later integration and psychological healing. The retrieval allows the patient to form a new and more specific narrative about his or her personal life history, a narrative often told for the first time.

The protocol also articulates the critical role of the therapist in providing cues to the retrieval process. Initially, access to specific autobiographical knowledge may be contingent upon first accessing general knowledge, as when the patient states, "She would often scold me for being slow". Autobiographical retrieval is thus quite different from access to semantic knowledge or access to episodic memories for single events, retrievals that do not require these efforts (Anderson & Conway, 1997).

When one wishes to remember experiences of the self in the past, a template must first be created by centrally located control processes. This retrieval template then modulates memory construction and, as in the example above, an iterative cueing cycle of search-evaluate-elaborate takes place. Thus, when the evaluation phase does not lead to the formation of a memory because the criteria in the template are not satisfied, the cue is used for further elaboration, as when the patient in our example specifically states, "My mother was quite abusive when I grew up."

When applied to the patient reliving an instance of childhood abuse, the same sequence is noticeable as in the Conway & Fthenaki protocol. In response to the analyst's cue, "How was your mother when you were a child?", the following occurs:"

1. Access to a particular period of life: "My mother went to work when I was seven."
2. Patient's elaboration of the cue: "My mother was quite abusive when I grew up."
3. Information common to lifetime period: "She would often scold me for being slow."
4. Access to particular period of life: "Eating my breakfast in the kitchen, where she always served it, on a little fold-out table near the door."
5. Access to sensory-perceptual details: "She suddenly came up behind me and kicked the chair I was sitting on so it fell over. I fell backwards and must have banged my head on the floor."
6. Other event-specific knowledge: "She just continued to call me names, told me what a slowpoke I was."

Because of its importance in maintaining a cohesive sense of self, Conway and Fthenaki call this protocol *generative retrieval* (2000, p. 304). When we wish to remember experiences of the self in the past, a template is created by what they call *centrally located control processes* (ibid.). This retrieval template then modulates memory construction and, as in the examples above, an iterative cueing cycle of search-evaluate-elaborate takes place. Thus, when the evaluation phase does not lead to the formation of a memory, because the criteria in the template are not satisfied, the cue is used for further elaboration, as when the patient in our example makes the comment "My mother was quite abuse when I grew up."

Levels of retrieval

This cueing process corresponds to what Conway (2002) describes as three levels of retrieval and how these levels depend on the amount

Table 3.1. Levels of retrieval and types of memory.

Knowledge areas	Conway research	Other research
Level 1	Lifetime periods	True episodic memory*
Level 2	General events	Generic event memory**
Level 3	Event-specific knowledge	Autobiographi-cal memory

* Tulving (2002); ** Nelson (1993).

of effort expended in remembering. Two of these levels have to do with memory that is more generic and requires less effort; these levels thus result in such abstract and conceptual knowledge as recalling a certain period in the rememberer's life or when some general events took place. Only the last method, requiring the most mental effort, involves a full retrieval of what the Conway group calls *event-specific knowledge* and *near-sensory experiences*—full autobiographical knowledge.

Based on levels of retrieval effort, the memory components are:

1. Knowledge that relates to *lifetime periods* (level 1): such as years spent in high school or having a particular employment. Lifetime periods therefore have a beginning and an end, as in such statements as "when I was in high school" and "when I was working at X" (Conway & Pleydell-Pearce, 2000).
2. Knowledge related to *general events* (level 2) that are still nested within lifetime periods: such as "eating in a restaurant" and "vacationing in Italy". General events are thus organised according to a common theme. They are often remembered because of their vividness, like going to a ball game for the first time, and may also relate to a goal, a success, or a failure, as in passing an exam or being turned down for promotion (ibid.).
3. *Event-specific knowledge* (level 3), or true autobiographical retrieval: these are vividly experienced visual records of near-sensory experiences. They have a high level of detail and tend to endure longer than the other two types of knowledge and they are often used to distinguish events that were actually experienced from those

imagined. However, event-specific knowledge can only be accessed by specific cues, such as those related to general events (ibid.).

In later describing this research, Conway (2004) states that autobiographical information usually enters our mind in the form of statements, propositions, declarations, and beliefs about the self, "accompanied by generic and/or specific images of details of prior experience" (p. 563). Thus, the retrieval of autobiographical memory appears to have a stable pattern of activation but one that depends on certain narrative constructs called *knowledge structures* as noted earlier by Conway & Fthenaki (2000, p. 304). These structures, in their linkage to episodic memories as specific sensual-perceptual memories of actual experiences, are what make the resulting memories uniquely autobiographical.

This schema differs from the Tulving (2002) classification described in the previous chapter, in which only two of these levels are examples of episodic memory involving autonoetic awareness. Strictly speaking, therefore, only the third level—event-specific knowledge—is true autobiographical memory. (See Table 3.1, above.)

What is significant about this third level of memory is that one of its components is a form of episodic memory that Conway (2004) characterises as "retention of spatio-temporal information (a hallmark of episodic memory)" (p. 564). Although this definition is narrower than the more common definition based on Tulving's work—according to which most memory for past events involving mental time travel is regarded as episodic—under Conway's definitions, episodic memory is an element in autobiographical memory and the two are related but not identical.

Thus, the first knowledge area in the Conway and Pleydell-Pearce model, dealing with *lifetime periods* (level 1), will in this book be called *episodic*, since it appears to require autonoetic awareness as a distinct state of mind. The second knowledge area in this model, *general events* (level 2), seems to cover both unique and singular events as well as a more generic form of memory representing repeated experiences of a similar event. As discussed in the previous chapter, generic memory for repeated experiences is unlikely to give access to specific events in the past (Nelson & Fivush, 2000).

The third knowledge area, *event-specific knowledge* (level 3), is therefore the only one that can be regarded as fully autobiographical. For this third knowledge area, the Conway team makes a good case for considering full autobiographical retrieval as distinctly different from episodic memory (Conway & Pleydell-Pearce, 2000; Conway, 2002). Here, true autobiographical retrieval no longer relates to a limited time period or memory for general information. It also serve a particular purpose: the integration of personal experiences over a longer time in order for the retrieved memories to produce a new sense of self.

Spacial and temporal cues

By definition, retrieval of both episodic and autobiographical memories is based on what many researchers call *recollection* (Rugg, Johnson, Park, & Uncapher, 2008; Yonelinas, 2002). In its simplest form, recollection is an explicit or conscious function; it involves remembering a unique event, such as where I parked my car today rather than yesterday. In order for this mental operation to be successful, I must retrieve information not only about the act of parking but the particular location where this took place and when it did. For the most current information to be available, however, the encoding of yesterday's location may be the stronger one; thus, only by also consulting a temporal cue will I arrive at the place where I parked my car today.

In itself, this is not an operation that involves any discrepancy between past and present self. The experience is uniquely mine and, at the same time, very common and generic. We may even question whether autonoetic awareness is required. What is involved is mostly my attention and my ability to focus on both spacial and temporal cues.

For at least two of the level of recollections that Martin Conway and Christopher Pleydell-Pearce (2000) outline above, a particular reflective state of mind appears necessary. Both lifetime periods (level 1)

Table 3.2. Levels of episodic memory.

Knowledge area	Method	Effort	Awareness	Narrative
Lifetime periods	Distance	Mild	Autonoetic	First-person scripts
General events	Location	Medium	Mixed*	Third-person scripts
Autobiography	Both	Strong	Near sensory	Self-narrative**

* See, Wheeler, Stuss, & Tulving (1997).
** A narrative equivalent of autobiographical knowledge. See Bruner (2002) in this book, Chapter Four.

and event-specific knowledge (level 3) often involve experiences from the distant past and thus may present discrepancies in experiences of self. Lifetime periods also turned out to be the most abstract of the three knowledge areas. When participants dealt with school-related experiences, for instance, generic images of teachers, classrooms, and thoughts about particular class topics usually surfaced. Evaluation of what the participant was good at in school versus not so good at, and what the person liked or disliked, were also common, as well as what he or she achieved academically. These evaluations would then mark the start and completion of a lifetime period, sometimes assessed as a difficult period in the person's life, sometimes as a positive one.

General events retrieval (level 2), on the other hand, was found to be particularly useful in remembering events of a more recent origin. Here clarity, accessibility, or the elaborateness of the memories can be weighed by judging their distance from current circumstances. When based on repeated experiences, retrieval of a general event may then lead to the formation of mini-stories, such as learning to drive a car or first romantic relationships (Conway, 2002). As mini-stories, however, general events retrieval will lose some of the stories' first-person characteristics, and when it does, the stories are most aptly described as *generic event memories*, or scripts. (See more detailed discussion in Chapter Four.)

Yet, in other instances, general events retrieval may serve as an entry point for more demanding retrieval. Conway writes:

> General events are, then, heterogeneous and contain information that can be used to access lifetime periods or sensory perceptual EMs [episodic memories]. It is these latter knowledge structures that form the third level of autobiographical knowledge ..., although it might be noted here that when an EM or set of EMs are included as an active part of a constructed memory they always evoke recollective experience (the sense of the self in the past).

(2002, p. 57)

In distinguishing between levels of retrieval, Conway and colleagues may at times touch upon the participation of unconscious and implicit elements in remembering certain past events. Especially, the *near-sensory experiences* they discuss may in some instances be accessed via such implicit routes, especially when it comes to highly emotionally charged memories (Siegel, 2003).

Therapy translations

In psychotherapy, restoring a sense of self usually means also reconstructing a sense of personal history. Thus, initiating a search for when and where important events in the patient's life occurred is often a significant portion of the treatment. A cueing cycle has to ensue, consisting of several steps of searching-evaluating-elaborating. The therapist's role in this cycle is as the facilitator, but also as the witness to a newly found experience of self.

In the case above, the cueing cycle was initiated when, as the therapist, I explored with the patient his experience of his mother when he was a child. This is a common exploration in an initial assessment and in Martin Conway and his group's research lifetime periods are the preferred level of processing since less of a cognitive effort is required (Conway, 2002). In this instance, the patient very quickly retrieved memories from a particular lifetime period, that of his mother's going to work when he was seven and the events just before it. His remembering that there was a time before she went to work, when she was home with him and his older brother, then led to his general events retrieval (level 2) in first person of eating breakfast in the kitchen.

After an intense pause, the patient was now able to access previously repressed memories of event-specific knowledge (level 3), but only after yet another cue from me, the therapist, when I asked: "Could you tell me what you remember about it?" Thus, the cueing cycle was kept alive by my reinforcing search cue, "How old do you think you were?", again with reference to a lifetime period; and at this point, the full emotional impact of the abuse began to be relived.

The autobiographical nature of the patient's experiences as a child would probably not have been available unless he had maintained in memory his declaration that his mother was quite abusive to him—thus, knowledge that he has maintained about his childhood. How this information is held in memory for such a long time—and in spite of the obvious dissociation from its emotional impact—is often impossible for the clinician to establish. From a neurocognitive perspective, however, the statement reflects a general events level of memory and was most likely due to the patient's repeated experiences of a similar nature. As research into dissociated trauma shows, without this general knowledge of what happened to him at age six, the patient probably would have not been able to access the specifics of what happened to him. He would have been left with what Bessel van der Kolk calls "sensory fragments that have no linguistic components" (1996, p. 289).

As Helen Williams, Martin Conway, and Gillian Cohen (2008) point out, general events memories are particular in that the field perspective may be switched. Basing their views on several studies, they conclude:[2]

> Changing from a field to an observer perspective had the effect of diminishing affect. The findings showed clearly that it was possible to switch perspective and most memories could be recalled in either the field or the observer mode, although it was harder to switch if the memory was old or not very vivid.

> (p. 23)

In this instance, we appear to deal with an observer perspective; thus, the statement reflects a reconstruction rather than a direct re-experience from the original viewpoint.

Cueing the therapist's memory

Unfortunately, most therapy formulations still labor under the outdated notion of memory as a unitary faculty and, as I discussed in the previous chapter, may place too much emphasis on conceptual and abstract approaches to past experiences—consequently, semantic knowledge. To be successful, therapeutic treatment needs to promote access to the patient's episodic and autobiographical memories. Such retrieval requires effort, but therapeutic treatments are well suited to creating an environment for the personal search and introspection associated with that kind of work. The patient's efforts are directed toward answering cues provided by the therapist or, in some instances, also by the therapeutic setting itself.[3] This process, in turn, encourages the patient to relate in first-person statements characteristic of episodic retrieval.

Being able to distinguish between different levels of retrieval allows the therapist a wide range of approaches to the patient's history. Both lifetime periods and general events contain information that can be used for further cueing, either to other lifetime periods or to direct sensory experiences and event-specific knowledge. Conway terms this process "one-to-one mapping in cues", a process that allows the patient with memories of abuse to remember in near-sensory terms.

As the Conway-Fthenaki (2000) protocol shows, lifetime periods are also the cues to the therapists' own episodic memories of earlier phases of a treatment. General events memories, on the other hand, are most often retrieved in an observer perspective as third-person scripts; hence, they are no longer limited specifically to a current therapy as a lived, personal experience. Instead, they are attached to an entire storehouse of mini-stories, thus are the result of many treatments in which the therapist participated, ongoing treatments as well as treatments that ended a long time ago.

Over years of practice and in making case presentations, these general events memories become stories that therapists tell, retell, arrange, and rearrange to fit particular cases. In the process, these stories lose their connection to specific episodic memories and become propositional statements and beliefs, often presented as "my experience tells me". As mini-stories or scripts, they represent certain conclusions derived via repeated experiences.

The type of memory therapists rely on from session to session, however, is most appropriately defined as lifetime periods, or simply *episodic*. Accessed via what Endel Tulving attributes to autonoetic awareness as a distinct state of mind, the therapist's memory for an ongoing treatment relies heavily on inherent contextual cues (2002). By being cued to something the patient discussed in a previous session, the therapist will be able to introduce a wider, more complete perspective that either confirms or contradicts what the patient is maintaining about himself or herself.

Neuroanatomy of retrieval

As we saw in the previous chapter, research has implicated sites in the hippocampus and related structures as critical to encoding and retrieving episodic memory. For instance, Joseph LeDoux (2002) treats sites in the posterior parietal cortex, the parahippocampal region, and areas of the prefrontal cortex as convergence zones for memory of the past. These zones enable critical cortical functions that mature late but result in increased connectivity between the medial temporal lobe and the prefrontal cortex (Nelson, de Haan, & Thomas, 2006). Via connections to more posterior areas, the prefrontal functions are related to language, attention, and motor control and also play a critical role in autonoetic consciousness (Wheeler, Stuss, & Tulving, 1997).

The strongest support for the idea that there is a specific memory that deals with the past comes from the study of patients with damage to brain regions involved in visual processing (posterior regions and occipital lobes). These patients were unable to generate images of the past and became amnesic. Since episodic memory traces are mainly encoded in visual images, amnesia results from this kind of brain damage (Conway, 2004; Hodges & Graham, 2002).

In studies of patients with lesions at other sites as well, it was found that specific memory of personal history was also disturbed (Conway, Pleydell-Pearce, & Whitecross, 2001). Conway and Fthenaki (2000) conclude that autobiographical memory appears to be a further elaboration of episodic retrieval and is distributed in networks in many areas of the neocortex. Since loss of access to autobiographical memory—rather than the information's being permanently unavailable—is the result of most retrograde amnesias from various frontal lobe lesions, they conclude that construction of autobiographical memory requires a more complex retrieval process than do simpler forms of memory about the past (ibid.).

In Conway and Fthenaki's research, full access to autobiographical memory is mediated by networks in the frontal lobes and by selectively and iteratively sampled knowledge networks in the temporal lobes. Through these networks, access is gained to "representations of ESK [event-specific knowledge] in posterior regions (occipital and parietal lobes)" (2000, p. 308). They conclude, therefore, that autobiographical memory has rather unique and complex retrieval routes.

Michael Gazzaniga of the University of California–Santa Barbara comes to a similar conclusion.[4] He writes:

> The picture that emerges is that aspects of self-knowledge are distributed throughout the cortex, a little here, a little there. There is some evidence that the frontal regions of the left hemisphere play a pivotal role in setting the goal for retrieval and reconstruction of autobiographical knowledge.
>
> (2008, p. 306)

Gazzaniga also holds out the possibility that recognition of familiar others relies on structures in the right hemisphere, while self-recognition "might be supported by additional left lateralized cognitive processes" (p. 307). Similarly, Eleanor Maguire (2002) of University College London, in reviewing results from neuroimaging studies, finds an overall pattern of medial left-lateralised activation associated with retrieval of autobiographical memory. Thus, the medial frontal cortex and left hippocampus appear particularly responsive to recollective experience.

The meaning of self

The frail nature of memory and its propensity for distortion has often been used to discredit the psychodynamic emphasis on regaining access to the past. Research findings certainly show us that past experiences are incomplete records, by no means to be equated with recording on a film medium (Loftus, 1981; Schacter, 1996). But the discovery of episodic memory—and its extension into autobiographical knowledge—is a clear argument against any broad conclusion that access to past experiences is impossible or has no significance for a person's psychological health.

Without access to the experienced past, even when such access has to be a reconstruction, the patient is left without meaningful tools for dealing with painful experiences that have led to stress and even dissociation. For this reason, the findings about episodic and autobiographical memory in many ways validate the psychodynamic emphasis on the significance of retrieving the history of a patient's past.

In practical terms, the complex activation of autobiographical memory also means that there are extensive demands on the brain's processing capacity. On the other hand, the knowledge base appears extremely sensitive to cues and does not always require conscious mediation. The retrieval, therefore, may occur while other processing sequences take place.

For the practising therapist, these conclusions do not seem surprising. The treatment setting is uniquely suited for the kind of "effortful" reconstruction of past memories found by memory researchers and it often takes time and repetition for such memories to be securely integrated. It is, nevertheless, remarkable how precisely these models pinpoint certain critical processes in a therapeutic treatment, when, for the longest time, the common wisdom had been that no such confirmation could be had (Spence, 1982). Translated to therapeutic treatments, we can now understand the common phenomenon of a patient's entering into a state of remembering autobiographical details from earlier phases of life.

Based on current understanding of the therapeutic process, the therapist may not be aware of having cued the patient's retrieval of memories. Existing psychodynamic models are simply not specific enough to account for the cueing events. The Conway-Fthenaki (2000) protocol will, therefore, be most helpful in working more precisely with the activation of memories of past experiences. In particular, the understanding that common knowledge structures may serve as entry points into the patient's retrieval process ought to facilitate the treatment of patients with early traumatic experiences. Critical to this work is also the understanding that cueing may have to be repeated before the retrieval can occur (Schacter, 1996).

Past self, present self

When patients begin remembering painful events in the past, their mood is often depressed even though the recollections are vivid and full of emotion. This was certainly the case for the patient (quoted above) who remembered particulars of his mother's abuse when he was a child. What had been repressed but carried in his long-term memory—and possibly caused his depression—now felt very disturbing to him. The image of himself as a young boy, he discovered, differed from his image of himself as an adult. How he understood his relationship to his mother as a child, no longer held true for him as an adult. He had, in fact, two images of self, one based on cognitions when he was a child, as his mother's loyal servant and companion, always eager to please; and the other, on cognitions of his adult self, as living in fear of her violent and unpredictable rages.

Autobiographical memory is, accordingly, an important base for how the person's sense of self is maintained—a sense of self that nevertheless undergoes many changes in the personal life cycle, both cognitively and emotionally. In the therapy setting, retrieving memories from childhood often causes a clash between a past and a current experience of self, and the emotional intensity of the retrieval may well be explained by this clash (Conway, 2001). Autobiographical memory, although vulnerable to dissociation, thus serves the function of maintaining and updating a coherent reference point in consciousness about one's sense of purpose, basic beliefs, and goals to be attained.

Early self formulations

It took fifty or more years for the concept of *self* to become part of the vocabulary in most psychodynamic paradigms. The original psycho-analytic conception included only an immediate sense of self as "I", or ego as it functions in the here and now (Breger, 1974; Fancher, 1993). Although patients by and large referred to themselves as a coherent whole, most therapists appeared disinterested in how this sense was maintained.

One of the results of introducing the concept of self was that a rela-tional perspective replaced the emphasis on intrapsychic dynamics in Freud's work. The relational therapist Stephen Mitchell summarises this new perspective as follows:

> We are concerned with both the creation and maintenance of a rela-tively stable, coherent sense of self out of the continual ebb and flow of perception and affect, and the creation and maintenance of dependable, sustaining connections with others, both in actuality and as internal process.

> (1988, p. 35)

So far, however, the new emphasis on the significance of maintaining a cohesive self are seldom linked to memory; in particular, to autobio-graphical memory. Only recently have therapists begun to grapple with this new data—data that, as I show in the second part of this book, will require some radically new formulations, especially in relation to thera-pists' reporting and training.

The rare exception in formulating theory about the self is C. G. Jung. As early as 1928, he articulated a concept of the self that is mainly based on visual imagery, or what he called *archetypal symbols* (1961a, pp. 211–212). According to this theory, the self stands for wholeness and individuation and is the goal of human development (1935b, p. 174). In his autobiography, he describes this realisation as follows:

> During those years, between 1918 and 1920, I began to understand that the goal of psychic development is the self. There is no linear evolution; there is only a circumambulation of the self. Uniform development exists, at most, only at the beginning; later, every-thing points towards the center. This insight gave me stability,

and gradually my inner peace returned. I knew that in finding the mandala as an expression of the self I had attained what was for me the ultimate. Perhaps someone else knows more, but not I.

(1961b, pp. 196–197)

For various historical reasons, this conception of the self has had a limited influence on other psychodynamic formulations, but it has remained a central concept to Jungian formulations of treatment.[5] As revealed by the recent publication of Jung's personal diaries leading up to his theories about the self, these hypotheses came about not from research findings—very little was known about the neurocognitive aspects at that time—but from personal explorations of his own dreams (Jung, 2009).[6]

Self and consciousness

The approach used by most neuroscientists differs from psychoanalytic formulations and avoids introspection: its aim is to establish formulations about the self that are based on available neuroscientific data about memory and its functions. However, when no clear neuroanatomical maps appear to surface, some neuroscientists use evolutionary speculations to fill the void. The most ambitious of those has been formulated by the neuroscientist Antonio Damasio of the University of California at San Diego and is still based on the self's being tied to an understanding of consciousness. Using an approach similar to Freud's and his drive theory, Damasio (2010) traces the self to the basic signs of life in a form of alertness that we share with most animal species. He writes:

> Brains begin building conscious minds not at the level of the cerebral cortex but rather at the level of the brain stem. Primordial feelings are not only the first images generated by the brain but also immediate manifestations of sentience. They are the protoself foundation for more complex levels of self.

(p. 22)

Damasio calls the self "a process, not a thing, and the process is present at all times when we are presumed to be conscious" (p. 8). Even this definition is problematic. After all, if the term *self* is interchangeable with the term *consciousness*, it covers not one but several states of mind, from an unreflective sense of being alive and alert to certain reflective and contemplative states. Moreover, even the mind active in dreaming represents a form of consciousness (Foulkes, 1999).

In Damasio's schema, the self is a gradual evolutionary development from a protoself (which he locates in "specific sectors of the upper brain stem, a set of nuclei in the region known as thalamus, and specific but widespread regions of the cerebral cortex") to a core self and, finally, to the autobiographical self (2010, p. 23). Damasio admits that this is a hypothetical answer to an age-old mystery but is willing to go beyond introspection, behavioural manifestations, and existing data about the brain to invoke a combination of evolutionary theory and neurobiology (p. 15). In using these broadly biological functions to support his theory, what he arrives at is in many ways a new metapsychology.

His theory inherits the problems found in similar broad approaches. Until we have data about specific neural mechanisms about self, we are on more solid footing when we stay with what is known about episodic memory and autonoetic awareness, as discussed in the previous chapter. As his neuroscience colleague V. S. Ramachandran (2004) of the University of California–San Diego, points out, we are still unsure what we actually mean by the *self*. Ramachandran lists five different characteristics—continuity, coherence, embodiment or ownership, agency, and capacity for reflection—and concludes:

> Any or all of these different aspects of self can be differentially disturbed in brain disease, which leads me to believe that the self comprises not just one thing, but many. Like "love" or "happiness", we use one word, "self", to lump together many different phenomena.

> (p. 97)

For these reasons, I favour formulations based on current findings about memory. One connotation of the self is as *object*: when, as observers, we notice how our mind works, how certain behaviours or traits can be ascribed to what we call "me" or "myself". The other self is as *knower*: how, via reflection and introspection, we claim that certain experiences belong to us personally. This is the first-person memory that Tulving's research has found to be based on autonoetic awareness.

This means that I have rejected the idea—more or less implicit when tying the self to consciousness—that there is only one form of consciousness and only one aspect to the self. Instead, I will qualify the term *self* by using terms such as *self-image, self-narrative, working self, dreaming self,* among others. By using this approach, I hope for the data to speak first, as much as this is ever possible. As Michael Gazzaniga points out, the self represents several types of knowledge, several types of awareness. To him, there are at least four such types (2008, pp. 301–302):

1. *The conceptual self.* This aspect consists of the rememberer's theory of how he/she got to be the person he/she is, and includes notions of social identity and moral status as well as the capacity for a theory of mind and for empathy.
2. *The narrative self.* Stories constructed and rehearsed about ourselves and told to others about our past, present, or future.
3. *Self-image.* Details about our face, body, and gestures.
4. *Personality traits.* A network of information about memories and experiences that is both episodic and semantically based.

For formulations about psychotherapy treatments, the first two types of knowledge, conceptual and narrative, are the most relevant. Conceptual knowledge relates specifically to how the sense of self develops. Influenced by developmental research, the current view is often to treat the therapeutic relationship as reflecting these early patterns of relating. Of equal relevance, although often missing in psychodynamic formulations, is the second type of knowledge, which is based on a narrative self. Neuroscientists have made great progress in pinpointing this as part of autobiographical memory. (See Chapter Four for the particular kind of stories that can be regarded as self-narratives, how they are constructed, and their direct connection to episodic memories.)

The last two types of knowledge—self-image and personal traits—are often discussed together under the rubric of self-image. Here, social perspectives dominate; and, for some authors, self is primarily a social construct. To Louis Cozolino, for instance, self "is a matrix of learning and memory organized and encoded within hidden layers of neural networks" (2002, p. 170). In this perspective, the self represents "our way of being, thinking, and feeling that supports our connection with the group" (p. 171).

The working self

When focusing on the working of the self in remembering past events, Martin Conway and his colleagues have arrived at yet another definition.[7] They propose that autobiographical knowledge—for lifetime knowledge as well as general events—combines dynamically with episodic memories via a control structure they call the *working self*. As an individualised sense of self, this concept of self is based on personal goals and autobiographical memories as ways of assessing these goals. Conway (2004) summarises how this self-concept functions within autobiographical knowledge:

> The working self consists of a currently active goal hierarchy (only parts of which are consciously accessible), abstract knowledge of the self, and other knowledge that facilitates access to autobiographical knowledge structures. It is through the working self that new autobiographical knowledge and episodic memories are formed (encoded) and the working self also influences memory construction by controlling input to the knowledge base and in evaluating output (activated autobiographical knowledge).
>
> (p. 563)

As a higher order of cognition, autobiographical memory may serve many other functions; but for Conway its most significant function is in grounding the self by placing "constraints on what goals the self can realistically maintain and pursue" (2002, p. 55). This implies that memory and the self strive to be congruent. Although it may appear as if the self were "an illusion our minds attempt to create", as suggested by researchers,[8] Conway (2002) insists that the working self "makes preferentially available memories and knowledge that are congruent with the goals of the self", thus the need for memory and self to be congruent may override knowledge that initially appears incongruent (p. 55). In this sense, attitudes and beliefs may change or memories may be altered, misremembered, or inhibited in order to preserve the self from change.

This also means that "when the self and memory become split and no longer constrain one and each other", delusional beliefs and a variety of mental disorders are related to such failed strivings to make memory and the self congruent (Conway, 2002, p. 56). To support this view, Conway refers to Alan Baddeley, a senior British researcher, and interviews with schizophrenics who claimed that their delusions were actual memories (Baddeley, Thornton, Chua, & McKenna, 1996). A patient's conviction that most of his brain had been removed, for instance, Conway views as having its own tragic consistency in that "memory no longer grounds the self in the sense that it places constraints on what the self can be" (2002, p. 56).

In describing the working self—its goals, ambitions, and striving for consistency—Conway does not mention narratives but appears to include them in the type of statements, propositions, and declarations about the self that accompany detailed, specific images of past experiences. He also suggests that in using lifetime periods (level 1) for assessing personal goals, these evaluations can be chunked into higher-order units and form what he calls *life-story schemas* (2002). Such schemas would relate specifically to the person's self-concept.

Other researchers have been more specific about these types of statements, holding that they are based on certain types of narratives that seem central to maintaining and accessing a sense of self. As discussed in the next chapter, narratives are critical to how past experiences are remembered, given meaning, and integrated into a personal sense of self. To Daniel Siegel (1999), this also means that there are implicit or unconscious elements in how a sense of self is maintained. Certain mental models and schemas (further discussed in Chapters Four and Six) have their origin in the child's early development and are likely to be intertwined with the explicit retrieval in the form of images. Siegel concludes:

> Narratives reveal how representations from one system can clearly intertwine with another. Thus the mental models of implicit memory help organize the themes of how the details of explicit autobiographical memory are expressed within a life story. Though we can never see mental models directly, their manifestation in narratives allows us to get a view of at least the shadow they cast on the output of other systems of the mind.

(p. 63)

According to Siegel, these mental models, part of the attachment experience, rely on brain structures that are intact at birth and play an important role throughout life; they can be located in the amygdala and other limbic regions for emotional memory, the basal ganglia and motor cortex for behavioural memory, and the perceptual cortices for perceptual memory (1999).

Narratives and integration

When one lacks such narrative references, memories of the past involving abuse and trauma will often remain dissociated and will not have an impact on the rememberer's autobiographical knowledge. In such instances, the patient has a vague sense of something's being explicitly remembered, but the retrieval lacks a clear source. The following example comes from Siegel's experience with a thirty-five-year-old woman who began recounting her childhood experiences with a violent and alcoholic father:

> When she began to tell her story, her eyes became filled with tears, her hands began to tremble, and she turned away from her therapist. She stopped speaking and seemed to become frozen, with a look of terror on her face. For the therapist, the feeling in the room was intense and consuming. The patient began to speak again, but this time spoke of her father's "positive attributes." Her nonverbal communication remained, though she wiped away her tears and tried to "compose" herself and not "worry so much about the past".
>
> (1999, p. 43)

Here, the disclosure of abuse cannot be completed. What follows the generic retrieval of several repeated events is accompanied by strong emotions, but the experiences cannot be verbalised. This unfortunate situation is most likely to occur when the therapist is unable to provide the appropriately safe environment and when the patient's fear of regression is too strong. Experiences previously encoded separately, both explicitly and implicitly, will remain unintegrated and the dissociation continues.

What makes painful memories conscious is, in other words, their association to a narrative that will sequence the remembered events so that dissociated self-states can be integrated (Bromberg, 1998; Siegel,

1999). Once conscious and explicit, otherwise wordless images can be verbalised and communicated to others. For the thirty-five-year-old patient, no such storyline seems to surface, and her experiences will remain unconscious until she has a way to tell her story. Only at that point will what she describes also involve her sense of self at the time when the abuse took place. By definition, such a narrative is explicit autobiographical memory.

Most therapists experience this type of failure in their practice and wonder what they might do differently. If we take into account the need to connect to a working self, there are some reasonable explanations that help us find our way forward. Autobiographical memory, once developed, serves the dual function of providing a sense of self while also constraining it. Thus, retrieval of autobiographical memories must be grounded in experience. Since the self defines who we are while at the same time limiting what we can be, it determines the goals we can hold and our aims in reaching those goals.

When the relationship between self and memory is impaired and autobiographical memory no longer serves this important dual function, fragmentation of the self is often the result. Severe "clouding" of autobiographical memories may, therefore, lead to what Conway calls "overgeneral memories" (2004, p. 564). In depression, for instance, patients may retrieve memories that lack details and become "much more schematic" (ibid.). Similar symptoms have also been observed in patients suffering from OCD (obsessive-compulsive disorder) and schizophrenia.

Treatments in a new light

"I remember one time I was eating my breakfast in the kitchen where she always served it, on a little fold-out table near the door." This vivid image of himself as a five- or six-year-old, being berated and physically abused, was how the thirty-seven-year-old patient responded to the therapist's question, "What was your mother like when you were a child?" Later in his treatment, as a new narrative about himself began to form, he would use terms such as "when I think of myself" or "the little boy in me" to convey how he now understood his history, how certain events still haunted him, and how those events explained why he behaved in a certain way in the present.

The process by which he began to remember what happened to him as a child reflects what Martin Conway and Airkaterina Fthenaki (2000) call a *generative retrieval* (p. 304). This search-evaluate-elaborate cycle creates a retrieval template "by centrally located control processes"; thus memory construction is modulated (ibid.).

The patient's access to his or her personal history, although essentially a reconstruction, is often a critical element in psychotherapy treatment. The research on autobiographical memory can now give the therapist a broad understanding of this process so that it no longer has to be exclusively intuitive. By supplying the cues, the therapist will facilitate the patient's gradual integration of retrieved memories with his or her working self.

Paradoxically, however, these memories, "effortfully constructed and effortfully maintained in consciousness", as Conway and Fthenaki (2000) suggest, cause adjustments to the self. At the same time, the self causes certain memories to be maintained and others to be forgotten shortly after they were experienced.

As an aspect of autobiographical memory, episodic memory also plays a central role in what the psychotherapist does. By being cued to his or her specific episodic memories for what occurred in earlier passages of a treatment, the therapist retrieves the personally experienced knowledge that is shared with the patient.

To understand how both types of memory—autobiographical and episodic—are held in memory, we are best served when focusing on their narrative components. Thus, in the next chapter, I examine the various narratives that are part of a psychotherapy treatment.

CHAPTER FOUR

Stories told and retold

Based on the findings in memory research discussed in the previous two chapters, I am proposing that psychodynamic treatments, in fact all psychotherapies, have a narrative base. Only by understanding the narratives of both therapist and patient—and how they are encoded, stored, and retrieved—are we able to account for how treatments often produce coherence and integration in the lives of patients. And only by understanding the narrative base of treatment will we be able to document what actually occurs without having to rely on paradigmatic and often outdated theoretical formulations. How this documentation is done is critical to the future of psychotherapy, to its credibility with the public, and to how we educate future practitioners to develop a full narrative competence.

My proposal is by no means unique. In a symposium on narratives in 1979, Roy Schafer of Columbia University's Psychoanalytic Center made early inroads into the narrative base for psychotherapy formulations and case descriptions, especially from an analytic point of view reframing many of the original psychoanalytic formulations. According to Schafer:

> People going through psychoanalysis—analysands—tell the analyst about themselves and others in the past and present. In making interpretations, the analyst retells these stories. In the retelling, certain features are accentuated while others are place in parenthesis; certain features are related to others in new ways or for the first time; some features are developed further, perhaps at great length.

> (Schafer, 1980, p. 31)

More recently, Daniel Siegel (UCLA Center for Culture, Brain, and Development) ties narratives to "a bilateral integrating process" (2003, p. 29). To him, "the autobiographical narrative process may be a fundamental part of cortical consolidation" and involves a number of regions, including the orbitofrontal cortex (p. 28). Thus, episodic and autobiographical memory is intimately connected to the narrative process in many forms of psychotherapy.

Without considering the neuroscientific findings about episodic and autobiographical memory, psychologists have also examined different examples of therapeutic narratives, from the demonic and tragic to the redemptive and traumatic (Alon & Omer, 2004; McAdams, 2006; Sorsoli, 2004). Amia Lieblich of Hebrew University of Jerusalem summarises the general perspective on psychotherapy that has emerged:

> Many forms of psychotherapy—from psychoanalysis to cognitive-behavioral therapy—involve life storytelling and retelling; in other words, they are based on narrative. The construction of a story in some form or shape is part of the intake or diagnostic procedures, as well as of the final report, every therapist creates, whether for herself, her peer practitioners, students, or agencies towards which she is responsible. It is also part of how the therapist mentally records each session with every patient as he or she makes note of "what happened" in the meeting.

> (2004, p. 4)

Stories and therapeutic process

There is today enough neuroscientific data to suggest that the role of narratives does not stop with how treatments are processed and reported on. The fact that psychotherapeutic processes consist of stories ought to be evident when examining what occurs already at the beginning of a treatment. The first story often starts with the therapist's asking something like, "So tell me, how did you hear about me?" and the patient's telling a story or what, in fact, is only the beginning of a story. Of course, all patient stories are different in some way, but they also have something in common, especially those at the beginning of treatment. Each patient's story is meant to answer the therapist's question, whether it was stated or not, as if to say, "The reason I am here, in this room, sitting across from you, a complete stranger, is … ."

Before a treatment comes to an end, the patient has told a seemingly endless number of stories—stories about other people, stories about the therapist, stories about recent as well as past experiences. The task of giving a coherent, liveable, and adequate narrative about one's own life is, in other words, quite complex. It requires effort, patience, and perseverance. What Jerome Bruner (2002), a pioneer in researching stories, calls *self-making* is a narrative art from the "inside" as well as from the "outside" (p. 65). The inside to Bruner is "memory, feelings, ideas, beliefs, subjectivity", while the outside is based on "the apparent esteem of others and on the myriad expectations that we early, even mindlessly, pick up from the culture in which we are immersed" (ibid.).

A story about oneself has to be told from both inside and outside. As such, self-making inevitably involves pleasing an audience. Bruner writes:

> Telling others about oneself is, then, no simple matter. It depends on what *we* think *they* ought to be like—or what selves in general ought to be like. Nor do our calculations end when we come to

Table 4.1. The patient's narratives.

Clinical description	Memory system	Narrative
Factual information	Semantic/explicit	General scripts
Initial disclosures	Episodic/explicit	Rehearsed stories
New disclosures	Emotional/implicit	Untold stories/trauma
Session related memory	Episodic/explicit	First person/mini-stories
Treatment memory	Autobiographical/ elaborate	Self-narrative

telling ourselves about ourselves. Our self-directed self-making narratives early come to express what we think others expect us to be like. Without much awareness of it, we develop a decorum for telling ourselves about ourselves: how to be frank with ourselves, how not to offend others.

(2002, p. 66)

Patients' initial stories express this wish to please and follow decorum. They have the desire to be interesting to the therapist, still very much a stranger, and to have his or her attention. The stories often reflect the way a child wishes to please its parents. After all, the first stories date back to when *storytelling* was first learned, under a parent's tutelage (Nelson, 1999).[1] They are stories meant to prove goodness and loyalty or stories demanding parental attention. They are also tales of failure, rejection, and despair, confessions meant to impress, not stories open for review.

In analytic terminology, telling childhood stories—those reflecting perceptions from when storytelling was first learned—is *transference* (Usher, 1993).[2] Roy Schafer points that, in using this term, the therapist explains how the patient is repetitively "reliving or reexperiencing the past in the present relationship with the analyst" (1983, p. 220). And in its ambiguity of roles and expectations, the therapeutic setting seems to promote such regression to childhood patterns of relating. Thus, the stranger in the other chair or behind the couch becomes endowed with the power to heal; and if a good rapport develops, a new self-narrative may eventually emerge. First, however, the old stories have to be tried.

Types of narratives

The term *narrative* covers a wide range of phenomena, from fiction and serious literature to historical accounts, legal and philosophical arguments, and case reports. Some narratives are difficult to identify as stories, and many narratives that develop in a treatment fit this category. At least in their initial form, they are most accurately classified as *scripts*, or scenes (Mandler, 1984; Schank, 1999). Other narratives—and here we have to include the patient narratives we have called *self-narratives*—are more easily identified as stories in that they are narrated in first person and give voice to real events and how they were experienced. They are forms of what Jerome Bruner, in the quote above, calls *self-making*.

Defining the term *narrative* from a neuropsychological point of view, Raymond Mar of the University of Toronto stresses the presence of a structure in which a series of actions and events unfold over time and "according to causal principles" (2004, p. 5). This definition is similar to the one quoted in Chapter One from Jonathan Culler (1997), the literary scholar. Mar also distinguishes between stories and expository texts, since the latter lack characters, elaborate depictions, and environments. Only in stories, according to Mar, are we witnesses to "the creation of an imagined world which mirrors our own realm of experience" (2004, p. 5). Some therapeutic narratives—the scripts and master narratives, described below—may, from this point of view, be more appropriately termed *expository*.

When discussing the narratives in psychotherapy treatments, I will make the following distinctions:

- *Decorum stories*: The initial and well-rehearsed stories that the patient tells in response to what he or she assumes is expected by the therapist
- *Self-narratives*: The narratives that the patient develops during, and as the result of, a successful treatment
- *Therapeutic scripts*: The therapist's initial scripts and propositions
- *Treatment narratives*: The narratives encoded and retrieved by the therapist during each treatment and established as part of a therapeutic alliance
- *Master narratives*: The narratives that the therapist relies on, especially when presenting completed or well-established treatments.

These narratives may have several versions and go through many revisions and condensations. On the patient's part, the initial stories are stories that have been told many times. As decorum stories, they are part of his or her self-presentation, and thus also reflect many past experiences of self in its numerous social contexts. Gradually, however, the patient begins to replace these audience-pleasing stories as a new self-narrative develops progressively from several mini-stories, which eventually are omitted in the overall story.

Similarly, the therapist's treatment narrative develops from easily accessed therapeutic scripts. Before being able to fully engage in a treatment, the therapist has to integrate information particular to each patient with these initial scripts and propositions.

As fully developed stories, the self-narratives and treatment narratives also have certain given elements in common that make them distinctly storylike. Even as these narratives go through many versions, they have to include:

- Episodic memory particular to the treatment as it unfolds. For the patient, the retrieval of episodic memories is tied to his or her autobiography; in the end, these memories become part of the patient's autobiographical knowledge and sense of self. For the therapist, what occurred on a ongoing basis in a treatment is accessed as episodic memory.
- Storylines, or plots, that organise the narratives according to a particular subject matter. For the patient, the storyline is directly tied to his/her autobiographical self and first-person knowledge (see discussion in Chapter Three). For the therapist, the storyline is connected to select scripts that are dramatised into a story about each patient's history, personality, and process of change.

Roy Schafer defines the self-narrative as "the story that there is a self to tell something to, a someone else serving as audience who is oneself or one's self" (1980, p. 31). Thus, as a narrative about the self, it is anchored in autobiographical memory. As discussed in the previous chapter, this type of memory is a complex form of knowledge involving what researchers have called the *working self* (Conway, 2004).

The therapist's narratives—what I call *treatment narratives* using a term from the psychotherapist Robert Winer (1994)—on the other hand, are as much *for* the treatment as they are *about* it. These unique stories would not be possible without the many scripts and interpretative devices that are honed by professional experience. Ultimately, however, these scripts will become secondary to stories specific to each patient and each treatment.

The case of Jane

When the old stories lose their lustre, as in the following case vignette from my own practice, a marked change usually occurs in the relationship. Patients' disclosures that were never previously made may now enter the dialogue; additionally, the patient may wish to engage the therapist in a range of conflicting emotions and thoughts. But the telling of a self-narrative will also require what Robert Stolorow (Southern

California Psychoanalytic Institute) and his colleagues call "the shaping of the reality by the therapist's organizing activity" (Stolorow, Brandchaft, & Atwood, 1987, p. 135). The patient will need the complete attention and presence of the therapist in order to fully articulate his or her story (Ulanov, 1982).

Let's look at an excerpt from the narrative of one of my patients, following a short vacation of mine.

Jane, forty-five years old and married, with two children, had also been on vacation with her family. She began her session by asking how my vacation was and then, seemingly with great reluctance, she began telling about an incident during hers.

> I don't know why I am telling you this, but, as I said, we had a terrible time. The girls were good [referring to her daughters, five and eleven], and we spent ten days going down the coast—but why am I telling you this? I wanted to know how your vacation was—but I guess you are not going to tell me.—So, when you are on a boat, everyone has something to do all the time, so we didn't have any arguments or anything and I let him [her husband of twelve years] give the orders the way he likes to. He likes it that way and he is actually a good sailor. But that is not what I wanted to tell you about. I must be boring you.—But imagine what he said, just a couple of nights after we came back, when I wanted to hang that new painting I bought. We spent a day in Provincetown; I really had some fun, walking down the street, going into all these quaint galleries, and there I found this painting that had exactly the right colours, the right light. So he said: "I live here too. I don't want you to hang that thing in our dining room. It is ugly and I don't want you to take anything down, either." So now, this wonderful paint-ing is just sitting there in my closet. I can't talk to him, never could, he is just becoming more and more that way. Just wanting control, control of everything. So he hasn't said anything to me since and I am not going to give him the satisfaction anymore. He is absolutely hopeless, nothing is happening between us any more.

In her story, Jane appears to start a more authentic self-narrative, cued by the feelings of abandonment that the therapist's vacation has trig-gered. What she reveals are the sad details of a marriage relationship for which she appears to have lost all hope. However, inserted into her story are questions about the therapist, his ability to understand her predicament, and his trustworthiness.

The story may elicit a variety of interpretations. It could be understood as a tale of an insensitive and controlling husband. This interpretation certainly makes sense if one is identifying with the patient; that is, relating to the anger and despair she seems to express. However, the experienced clinician may also suspect that the verbalisation of this interpretation may be met with denial, even hostility, by the patient, and that there are other, more implicit, feelings involved. After all, telling the story seems to be her second choice after having failed to make the therapist tell about his vacation. We certainly cannot be sure if she wishes to discuss her marital problems or, rather, have the therapist speak about his vacation.

The listener could take a dislike to the woman telling her story. The tale about her sailing trip is a rather reluctant admission of a miserable vacation and she may seem to pretend not to have any share in the marital problems, thus seeming to lack a desire to be authentic. And, of course, the tale could be understood as revealing how insensitive most therapists are to their patients by not being willing to tell about themselves and hiding behind their analytic stance.

All these interpretations are possible; but a great deal more could also be discovered. If we assume that the forty-five-year-old woman wishes to be engaged in further exploration of what she told, we could imagine the story as a stepping-off place for further investigations about how she really feels about her marriage and why she is in therapy in the first place. Thus, the beginning of her self-narrative.

Ultimately, the interpretation arrived at is coloured by the kind of understanding the listener brings. Some therapists may be convinced that the story is yet another way to get the therapist to fulfil the broken promise of the patient's marriage, perhaps by reliving a childhood triangular drama to win the father's exclusive attention. Others may find the patient is repeating with the therapist a childhood pattern of trying to engage a distant, unavailable mother. In the end, however, the patient is cast in the role of the storyteller—even when, as is the case here, the patient relates the story rather reluctantly.

Creating new stories

When well-rehearsed old stories have exhausted their appeal, new ones have to be created. As a treatment like Jane's progresses to this point, both therapist and patient have probably concluded that there is a great deal of material still to reveal (Ulanov, 1982). But since humans, to a great extent, rely on knowledge and belief already learned, telling old stories is easier than creating new ones. As Roger Schank (1990) of Northwestern University points out in his discussion of stories as memory devices, creating new stories involves more of an effort than using old ones. Only by revealing previously untold experiences will we memorise entirely different stories.

When telling someone a story, we are telling a story to ourselves as well. We are, as Roy Schafer notes, "enclosing one story within another" (1980, p. 31). The story retained as memory is being reused and placed in new contexts, in combination with other stories told to self, about self.

For this to occur, patients must have concluded that old stories no longer offer a fit that is good enough. They must also have decided that they need to convey important new or untold experiences. So even if a patient's first set of stories are met with appreciation by the therapist, telling new stories is a necessary condition for psychological change to occur. Schank captures this fact in the following way:

> *We need to tell someone else a story that describes our experience because the process of creating the story also creates the memory structure that will contain the gist of the story for the rest of our lives.* It seems odd, at first, that this should be true. Certainly, psychologists have known for years that rehearsal helps memory. But telling a story isn't rehearsal, it is creation.

(1990, p. 115; italics in original)

In this perspective, what occurs for the patient in a successful therapy is not some form of rehearsal of already encoded material. The therapeutic process depends on the creation of a cohesive new story from many disconnected ones, a story that can span a longer time and encompass previously unintegrated experiences. Inevitably, this means having to be in the uncomfortable role of creating a new story while many of the old ones are augmented and revised to fit into a new structure (Josselson, 2004). The patient is asked to develop a new capacity to search and to explore often-painful memories, in stories initially seen as unrelated or nonexistent. Some of this material will relate to traumatic events, most likely encoded separately in what researchers describe as "cortical convergence onto the neostriatum and the amygdala" (Eichenbaum & Bodkin, 2000, pp. 197–198).

If a patient has no ability to remember and make these connections, his or her sense of self never becomes fully embodied (Covington, 1995). Integration and meaning thus come from the placement of the actors in the patient's autobiography within an overall configuration—the self-narrative (Bruner, 2002). This narrative will then include autonoetic experiences of the kind described in Chapter Two, when the fifty-two-year-old patient remembers being a teenager. Its subject matter is the patient's self, but it is also a story about the treatment in that it has a distinctly therapeutic origin. And since the latest encoding of previously unintegrated engrams took place in the treatment, the narrative will now contain certain retrieval cues—cues that may come into play when these memories are further consolidated. As a result, retrieval cues will give the patient access to critical aspects of the treatment long after it ends (Josselson, 2004).

Ultimately, the retrieved past experiences also have to be placed in sequence and related in time (Wilkinson, 2010). Certain episodes and events, explored over the duration of the treatment, will have to be omitted. Only those that adhere to the subject matter and certain causal principles will be included (Mar, 2003).[3] What during the treatment consisted of several stories, as instalments in a drama series with the same main characters, will be held together by the patient's increasing sense of being the teller of his or her own life.

Research on narratives

Due to the complexity involved in narratives and the great variety of storylines, researchers have found it difficult to determine precisely how they are encoded and retrieved. Nevertheless, narratives have offered surprising insights into conditions related to psychological dissociation and problems with attachment and intimacy (Hesse, Main, Abrams, & Rifkin, 2003). Ways of studying these conditions and how they can be remedied are now possible, in part by rating the narratives of test subjects.

As we saw in Chapter Two, the frontal lobes are critical to the retrieval of episodic material; thus, activation in these areas would be a necessary component in any patient's self-narrative and in the therapist's fully developed treatment narrative (Tulving &

Markowitsch, 1998). Other networks, such as those associated with imagery, comprehension, and integration, are clearly implicated (Wilkinson, 2010).

Anatomically, the two cortical hemispheres are somewhat different. As Siegel puts it, the right side is more "cross modal" and has more horizontal linkage. On the left side, "the cortical columns appear to work more on their own" (2007, p. 45). These differences make the right hemispheric cortical areas more oriented toward context and the whole picture, while the left hemisphere focuses on monitoring details and analysis (ibid.). But the two sides develop at different times in the infant, with the right hemisphere seeing substantial development during the first two or three years of life; left-brain functions mature only by the second year. Thus memory that depends on language may not be available for processing and certain traumatic experiences may only be accessible by nonverbal means.

Although he finds a considerable gap between cognitive models and neuroscientific data about narratives, Raymond Mar (2004) concludes that "right hemisphere activations were prominent throughout numerous studies" (p. 25). He also asserts that story comprehension (understanding others' stories) and story production (telling one's own stories) share similar patterns of activation, although production appears more complex in that it also requires semantic selection during generation; that is, when a story is verbalised (ibid.). Broadly speaking, however, any network involved in language, memory, and perception is likely to play a role; and networks associated with working memory must also be involved for text comprehension.

Quoting E. T. Rolls (2000), Mar points to the frontal lobe as critical for maintaining material in long-term memory, and he also specifies "neurons in the dorsolateral prefrontal cortex (specifically Brodmann's Area [Bas] 6, 8, and 9/46)", since these areas "have been associated with cross-temporal and cross-modal processing likely necessary for language processing" (2004, pp. 7–8).

Trauma and narrative

Based on neuroimaging studies, Daniel Siegel (1999, 2003) describes how narratives, especially those reflecting autobiographical memory, assist in overcoming splitting of consciousness into of what he calls *self-states* (p. 8).[4]

To Siegel and others studying psychological trauma, such split-off selves are due to dissociative processes in which emotions such as anger or shame may become engrained over time and form their own patterns of activity (Bromberg, 1998; Wachtel, 2008). However, narratives may help in establishing "a functional flow in the states of mind across time", a process Siegel terms *integration*:

> Autobiographical narratives can reveal integration or incoherence. A coherent narrative reveals a blending of left- and right-hemisphere processes. The interpreting left hemisphere is driven to weave a tale of what it knows. When access to the right hemisphere's representational processes is limited, such a tale is incoherent.
>
> (Siegel, 1999, p. 336)

Siegel bases this explanation on the fact that laterality—the specialisation of brain function in two cortical hemispheres—results in the necessity for the articulation of a story in the left hemisphere to combine with more implicit and broader contents in the right cortical areas:

> The linear telling of a story is driven by the left hemisphere. In order to be autobiographical, the left side must connect with the subjective emotional self-experience that is stored in the right hemisphere. The proposal is this: to have a coherent story, the drive of the left to tell a logical story must draw on the information from the right. If there is blockage, as occurs in PTSD (posttraumatic stress disorder), then the narrative may be incoherent.
>
> (2003, p. 15)

Attachment and narrative

Referring to several recent studies, Louis Cozolino (2006) and Daniel Siegel (2007) also conclude that the experiences one has before being able to form narratives may impact many aspects of functioning well into adult life.[5] Affective style and the way a person approaches interpersonal relationships appears to form between infant and caregiver at an early stage.

These studies, based on attachment theory as first formulated by John Bowlby and Mary Ainsworth, implicate the type narrative a person will articulate about himself or herself in the Adult Attachment Interview or AAI (Main & Goldwyn, 1998). And when these findings were compared to tests conducted on infants in the Infant Strange Situation (ISS) study developed by Ainsworth and colleagues, the AAI proved to be a good predictor of the attachment patterns transferred from caregiver to their children (Ainsworth, Blehar, Waters, & Wall, 1978). Aside from reflecting a secure attachment, the narratives from tested adults fell into the following patterns:

1. *Avoidant or dismissive attachment*, the first pattern in these studies, was found in the type of narrative these caregivers were able to produce about themselves autobiographically:

   > The narrative has a tightly wound cohesion that excludes relationships and emotions from being considered important now or in the past. This narrative organization is often marked by the insistence that the individual cannot recall memories of family experiences in detail.
   >
   > (Siegel, 2007, p. 202)

Louis Cozolino (2006) describes these findings in a similar way:

   > The narratives of these mothers were incoherent, usually due to significant gaps of time and information. Dismissing parents seem to cope through denial and repression to a degree that interferes with adequate integration of cognitive and emotional processing. The lack of recall suggests encoding deficits due to trauma, chronic stress, or lack of assistance in regulating affect and reinforcing memory early in life.
   >
   > (p. 145)

2. *Enmeshed-ambivalent or ambivalent-anxious attachment*, the second pattern, was found in caregivers who also had incoherent narratives, but this time with "and excess of verbal output" (Cozolino, 2006, p. 145). In Cozolino's words:

   > Enmeshed-ambivalent mothers demonstrated an excess of verbal output which they had difficulty organizing. They were also lacking boundaries between past and present events, further confusing

the listener. These mothers appeared preoccupied, pressured, and had a harder time keeping the listener in mind.

(ibid.)

3. *Disorganised attachment*, the category of the third group of caregivers, was seen in sufferers of unresolved losses and traumatic histories. The coherence of their narratives, according to Cozolino, "was disrupted by emotional intrusions and missing information about trauma and loss" (2006, p. 145):

> Their narratives not only reflected the disorganization of verbal and emotional expression, but also the devastating impact their unresolved traumas had had on their development and integration of the networks of their social brain.

(ibid.)

As parents, these test subjects manifested both terrified and terrifying behaviours toward their children, who often found themselves having the impulse to move away, while at the same time having the wish "to move towards the attachment figure for protection and soothing" (Siegel, 2003, p. 203).

The creation of a coherent autobiographical narrative thus serves a crucial role in developing secure relationships and has ramifications for treating these problems in psychotherapy. Whether the patient entered into treatment with narratives indicating avoidance of dependence, narratives lacking boundaries between past and present, or narratives reflecting intrusions and missing information, the process of co-creating a new self-narrative with the therapist is at the core of forming healthy attachments.

The therapist's stories

A necessary requirement for the patient's development of a new and coherent narrative is being in a relationship based on trust. The patient's story can only happen in a safe relational context. As a self-narrative, it has to be told—if not to the therapist, then to some other audience, or at least a confidant or two. In a successful treatment, what the therapist brings to the relationship is this personal context, as the engaged witness and trained participant.

In order for therapists to offer this relationship, they need to retrieve a set of tacit stories of their own, which in turn must be tailored to each treatment, each patient. In many ways, these treatment narratives are what allow a therapeutic process to take place—by providing the environment in which the patient's narrative can emerge. Without a therapist's narratives, the encounter will most likely come to an early conclusion (Nelson, 1993).[6]

Primarily in the listener's role, therapists rarely verbalise their stories from session to session. Initially, their stories are scripts more than fully developed stories and are based on a wide range of propositions, predictions, and beliefs (Bruner, 1991; Mandler, 1984). A patient may describe experiencing lack of sleep and appetite, while also complaining about loss of interest in everyday activities. In response, the therapist will, without much deliberation, take notice that the patient is depressed. Another patient may tell of persistent arguments with a partner, at the same time insisting that the partner is to blame for the problems. In listening, the therapist will focus on the patient's emotions and lack of self-reflection. While also hearing the anger expressed, the therapist will be attuned to the patient's pain of failure in the relationship. And when a patient, after just a few sessions, manifests rage over the therapist's perceived shortcomings, he or she will relate to the patient's painful wounding rather than becoming defensive.

Table 4.2. The therapist's narratives.

Clinical description	Memory system	Narrative
Clinical knowledge	Semantic/explicit	General scripts
Session related memory	Episodic/explicit	Mini-stories/first person
Therapeutic techniques	Procedural/implicit	None
Interpretations	Generic	Propositional scripts
Theory formulations	Semantic/explicit	Master/expository
Case related memory	Episodic/explicit	Observer narrative

Therapeutic scripts

In all three instances, we are dealing with what I term *therapeutic scripts*, implicitly encoded sources of information acquired by having undergone an experience many times. To Roger Schank, who first introduced the concept, a script represents "a very specific set of sequential facts about a very specific situation" (1999, p. 11). Therapeutic scripts thus provide the therapist with a range of predictions and hypothetical interpretations. In some instances, they are semantically encoded propositions and follow the simple logic of "if both x and y, then z" (Flew, 1979). In other instances, they are based more on intuition and beliefs that require no particular grounding in facts.

As discussed in the first chapter, scripts are early acquisitions, developed before the child has the ability to form episodic memories. In the words of developmental psychologist Robyn Fivush, they represent "what happens each and every time the event occurs" (1997, p. 142). In this sense, scripts are not abstractions of an explicit and semantic nature, which could be applied more broadly. And they do not give access to a memory of specific episodes in the past. Rather, they allow for a rather intuitive processing and comprehension, while inhibiting the specifics of each individual event from being encoded into memory (Hudson & Nelson, 1986).

Some of these scripts may have a rather short conscious life in a treatment, while others will be reinforced by the patient's disclosures. Some of the therapist's scripts may lead to interpretations, whereas others will remain ways of processing the patient's information internally. In contrast to fully developed stories, however, scripts often lack certain elements, such as complete storylines, and their sources are not accessible without specific probing.

Ruthleen Josselson (2004), in her case report "On Becoming the Narrator of One's Own Life", describes one such therapeutic script as "*my* master narrative". She writes:

> Therapy exposed Heidi to *my* master narrative that values inner experience and felt connection to others. I refrained from interpretation, recognising that I ran the risk of offering her linkage that she could not make on her own—which would thereby repeat her mother's behavior. Instead, I asked her lots of questions about her inner experience, her feelings, and her wishes. Mostly, this

perplexed her, but she tried, in her persistent compliant fashion, to produce something to please me.

(pp. 117–118; italics in original)

Without taking notice of the personal origin of her narrative, Josselson acknowledges its importance to the entire treatment. She also describes how, at some point in the treatment, her narrative has to include intimately personal information about her patient, as well as the reasons given for seeking treatment. This information is, in fact, what forms the basis for Josselson's particular approach to Heidi's process of healing.

The therapeutic alliance

How therapeutic scripts and patient information come together as a treatment strategy is determined at the early stages of the treatment, when the therapist-patient relationship, often called the *therapeutic alliance*, is being established (Safran & Muran, 2000). As a result of this alliance, therapists have a distinct sense of the patient's personality, critical experiences, and potential healing. This, in turn, often coincides with the therapist's having arrived at a narrative that is particular to the treatment.

The forming of a therapeutic alliance also tends to coincide with a patient's anxiety about the treatment situation's being replaced by the patient's sense of being understood (Wolf, 1988). As several outcome studies in psychotherapy indicate, a positive aspect in this alliance is a critical sense of collaboration (Horvath & Bedi, 2002).[7]

This sense may, in fact, be related to what the Boston Change Process Study Group (BCPSG) calls the *moment of meeting*, an emergent and intersubjective state in which the relational environment in an analytic relationship changes (2002). In their examination of moment-to-moment interactions between therapist and patient, they find that this alliance also signals the arrival of what they call *something more*, in which two minds acting together create something new and "what comes into being did not exist before and could not be fully predicted by either partner" (2005, p. 700).

The study group bases this conceptualisation on what it terms *implicit relational knowing*—relational patterns established in early infancy, which tend to endure throughout life. In a similar observation,

Jung describes this phenomenon as "a certain moment, sometimes long delayed when the foregoing preliminary talk *touches* the unconscious and establishes the unconscious identity of doctor and patient [italics in original]" (1945, pp. 182–183).

Most of the therapist's engagement in establishing this alliance occurs without his having to initiate a conscious search of memory. There is no thought like, *"Now I must find the proper story for this patient."* Rather, the therapist may experience an increase in attention and feel challenged to listen very carefully. The therapist's participation in a treatment, accordingly, begins with wishing to remember, to hold onto the information, verbal as well as nonverbal, that the patient supplies. And in order to do so, the therapist has to rely on stories that have been encoded into his memory at some previous occasion, in an entirely different context.

Diagnosis and narratives

Therapists with an orientation toward the medical model of practice will establish a diagnosis as part of the early phases of a treatment. In so doing, they usually use some procedure. With the diagnosis will also follow a certain story, which most likely is more implicit than the list of symptoms and their criteria required for a diagnosis. For instance, the diagnosis of borderline personality disorder was, for a long time, popular with many clinicians; in discussing this diagnosis with other clinicians, therapists often volunteered stories of a particular borderline patient they had treated. The diagnosis had become a remembered story, not merely a checklist of symptoms.

Even when the therapist professionally communicates his diagnostic assessment of the patient's problems, the therapist needs a story that has a beginning and an end, a story rooted in the patient's distinct and unique history, a story that shows how someone or something was transformed. Whether the story is tragic, redemptive, or otherwise, a storyline is needed. As a therapeutic alliance becomes established, the therapist's participation largely depends on having appropriate stories at his or her disposal.

This process may take several sessions. During this time, the therapist may develop an unconscious aversion to the patient and feel unable to understand what is happening. The situation of listening to the new patient and having many—often contradictory—reactions to what the patient says nevertheless forces this search for a story to continue.

Since no two stories are exactly alike, therapists may rely on the retrieval of stories with similar features (Schank, 1999). If reminded of a story, a case, or an experience that is similar enough, the therapist will have a sense of connecting to the patient, although the source of the memories—the original story, so to speak—is now beyond immediate access. In the process, a label or index has to be added to the original story so that the new variant can be included for future use and for when the therapist is faced with a similar experience (ibid.). Although learning has occurred, the therapist did not have to memorise the entire new story just heard.

The therapist may also feel that she is hearing something for the first time—that the expected sense of familiarity is not occurring. In an attempt to explain why the expected did not occur, an existing story has to be revised and expanded. The therapist may, in fact, have to use certain semantic knowledge involving propositional representations and symbols, such as when she encounters a mental disorder that she has had no previous experience treating. Again, new learning has occurred without her memorising the patient's entire story.

All of this happens while the therapist is in a state of heightened attention, and with great effort on her part. While waiting for an appropriate story to emerge, the therapist will probably have problems remembering much of what the patient says. Her attention is taken up by the search for some type of cue to her previous experiences.

Should the narrative fail to give the patient a sense of being understood, a successful therapeutic outcome is unlikely. We must then assume that the therapist is relying too heavily on preexisting scripts—she is not reminded of stories that feel applicable to the patient. When the therapist is unable to offer a genuine new understanding, the patient's early anxiety cannot be transformed into a feeling of being understood. No new narrative occurs with which to view past problems and no new tools emerge for dealing with those problems.

Narrating treatments

To be meaningful in a particular patient's treatment, the therapist's narrative must therefore hold critical memory about the information that surfaces throughout the treatment. Without explicitly disclosing the progress of this narrative, the therapist will rely on what it tells about the treatment. As Josselson (2004) shows in her case report, this information is essentially retained by the therapist as episodic memory specific to each treatment, a type of memory I call *session-related memory*.[8] Unless these memory traces are well encoded, chances are that the therapist will not develop a real rapport with the patient and that the patient will not feel understood.

This means that scripts and propositions from the initial assessment phase must be replaced by what Dan McAdams in his research on narratives calls "the ability to think in stories" (2006, p. 79). The therapist must construct a narrative with a distinct set of characters, usually reflecting the patient's original family constellation and relevant past experiences. These scenes must also be organised as enacted over time, from one scene to the next, thus providing a sense of the patient's history. At the same time, the resulting narrative has to consist of specific episodic memory of what occurs in the treatment. Once firmly stored in the context of a particular relationship, these memory traces can then be retrieved in future sessions, especially if the therapist has been able to place this information in a suitable narrative.

Retrieval of session-related memory thus serves a broader purpose. The therapist's narrative is about particularly relevant relational events of a memorable nature (McAdams, 2006). This also means that the material can be cued, but only contextually, when the therapist again experiences the situation in which the encoding occurred (Brown & Craik, 2000).

Although initially consisting of certain scripts that reference previous experiences with other patients, successful treatments depend on the therapist's ability to encode and retrieve enough relevant episodic memory particular to each treatment. Treatment narratives are, in fact, tailored to fit a unique circumstance—that of the individual patient, the individual therapist, and a uniquely developing therapeutic process. It is the record in narrative form that each therapist develops of the history of his or her relationship with a patient.

In the clinical literature, there are still few comparable discussions of how therapists process information particular to each treatment. Most formulations are meant to establish general rules. Rules, however, do not account for the fact that treatments, first and foremost, are dialogical and mostly consist of first-person statements in story form.

Most psychotherapy formulations assume that therapists approach treatments based on certain paradigmatic techniques. The individual therapist's episodic memory and previous learning, based on what has occurred in the treatment, are thus viewed to be of secondary relevance to the outcome.

When patients give voice to events they participated in, either in ongoing life situations or in the distant past, and try to reveal the impact of those experiences, they do so by telling stories. Although the therapist may be able to detect certain general storylines in these disclosures, the patient is communicating uniquely personal information. After all, the patient is the story's main protagonist. How each therapist structures this information to suit his or her particular way of remembering can only be fully explained when we understand the different components in treatment narratives: how they develop from initial hypothetical propositions into firmly encoded episodic memories of critical aspects of previous sessions.

The observer storyline

The cueing of memory becomes much more difficult when the therapist, outside of the immediate context of sessions in the consulting room, gives an account of a treatment. Especially when the therapist's purpose is to report on an entire treatment—whether still ongoing or terminated—a great deal of the episodic material is beyond access; further, some material has been permanently lost. The first-person nature of the remaining episodic traces also tends to change. The therapist is no longer one of the story's protagonists, but has become the story's narrator. The treatment has been placed within an *observer storyline* (Rice & Rubin, 2011; Williams, Conway, & Cohen, 2008).

As such, this storyline has to comply with certain requirements. It must confirm or reject the therapist's initial hypotheses about the patient. Which hypotheses are reinforced—and which become irrelevant—now seems to depend exclusively on how the patient responded; that is, on the patient's actions and disclosures. The therapist's narrative has become a treatment narrative and much of the therapist's ongoing but silent involvement in shaping the therapeutic interaction is left out.

This curious circumstance is related to a distinct and often overlooked aspect of fully developed stories: in order to tell what happened over an extended time period, when a great deal of memory has become inert, we rely on whatever story has been encoded into our memory. We need a storyline.

This is where an observer storyline comes into play, as an often-effective way to present historical facts without revealing the observer's participation in the collection and selection of the data. However, in placing the data in this context, there is a considerable risk of what research on memory distortions calls *hindsight bias*. The temptation is to stick to the original prediction, especially if the teller of the story—in this case, the therapist—predicted an outcome that later was contradicted (Metcalfe, 2000).

Even when the therapist wishes to present what occurred in a specific session, these implicitly stored storylines will eventually guide the story. The therapist will remember particular exchanges and disclosures as part of a story with a plot. This gives the account its sequential order and places the patient's disclosures in a time frame. But this is not all. The storyline imposes certain rules on the story outcome. "There must be an initial situation, a change involving some sort of reversal", according to Jonathan Culler of Cornell University, "and a resolution that marks the change as significant" (1997, p. 85).

This may not be clear to the therapist from session to session, but when presenting a treatment, in writing or just verbally, the element of transformation—an end relating back to the beginning—becomes part of the therapeutic narrative. Even when the treatment process had many other complicated turns and reversals, the storyline will preside over what is being included or left out.

Presenting what appears to be a straight-forward account of a therapy session thus involves more than the particulars remembered from the session. Well before the therapist's recall is translated into a formal account, certain storylines have been activated and, as I will discuss below, these storylines are an inevitable aspect of the therapist's participation in a treatment. The story-within-the-story comprises the necessary underpinning for a treatment narrative to form.

Storylines are, therefore, the independent underpinning that existed in memory prior to and independent of the specific scenes and images presented. As a level of structure "independent of any particular language or representational medium"—again quoting Jonathan Culler—there may be a seemingly endless number of such storylines. They are what give a story its narrative foundation and particular meaning (1997, p. 85).

These storylines may suggest a plot with certain parallels, such as the move from one type of relationship between the story's characters to its opposite (a different understanding of family relationships, for instance). They may also suggest the move from surmounting overwhelming fear to confirming or rejecting a certain outcome (e.g., moving from a problem with intimacy to forming lasting relationships), or from a prediction to its realisation or inversion (e.g., the processing of failure scenarios or doomsday prophecies).

Retrieval by indexing

To explain memory retrieval in everyday life, researchers in social cognition have pointed to *indexing* as one common option (Baumeister & Newman, 1995). Especially when it comes to scriptlike material, adding shortcuts or labels each time a story is modified makes the information broader and easier to apply to a wider cluster of situations. As secondary elaborations, these indexes may be of many kinds, and they function more or less as cues to already encoded narratives.[9]

One such index is beliefs. As juxtapositions to another person's beliefs, these indexes are a common way of extracting information from our own stories while listening to others (Schank, 1999). In everyday conversations, this allows us to enter into an exchange of opinions by introducing our own viewpoint, perhaps engage in a debate. In analytic treatment, this type of easy access is often of limited use. Indexing tends to prohibit further exploration of what the patient wishes to reveal. Instead of relying on labelled information, therapists—with some exceptions—train themselves to set aside references to their own previously indexed story material. In order to be effective in the patient's explorations of autobiography, they avoid communication in which topics are debated or viewpoints discussed (Safran & Muran, 2000). As far as possible, therapists instead strive to retain a continuous narrative of the information that the patient discloses. Thus, therapists build on narratives that hold a sufficient amount of episodic material about what has already happened in the treatment. As we saw earlier, this is particularly difficult in the initial phases of a treatment, before enough information has been disclosed.

The following case vignette illustrates this point. While listening to a patient describe his struggle to overcome dependence on painkillers, I was struck by how irresponsibly hospitals give prescriptions for these types of drugs. Upon being released after surgery for a broken arm, the patient, an alcoholic who had maintained sobriety for many years, was given a two-month supply of Percocet®, which he soon began to abuse. This reminded me of a time when I worked for a state agency for homeless men and was told that the agency sometimes gave out food vouchers that often were traded for cash, although the recipient may have been an active alcoholic.

Unfortunately, the story I retrieved was not particularly related to the patient whom I saw only for a second time. But since I had no personal experience of addiction to painkillers, I remembered how institutional policies sometimes can be misguided. In my responses to the patient, I therefore expressed disapproval of the hospital's actions and was unable to focus on his struggle to overcome the Percocet® addiction. Only when the patient reminded me why he brought up his problem with the painkillers was I able to regroup. Instead of relying exclusively on the indexed story from my experience in a social agency, with the therapy relationship as context and focus, I now remembered our previous discussion about his alcohol abuse and how he was able to overcome it.

A far more reliable approach when responding to a patient's stories are the therapeutic scripts described earlier in this chapter. These scripts consist of a wide range of propositions, predictions, and beliefs and will often elicit further disclosures (Bruner, 1991).[10] As the treatment progresses from session to session, it will be based on both participants' encoded episodic memories about the therapeutic relationship, memories that reference similar events. As in the case involving dependence on painkillers, even if I had had no personal experience of being severely addicted, my responses to the patient had to be based on empathic, first-person knowledge. This knowledge was more significant than access to stories based on indexed explanations from previous social circumstances.

While retrieval by indexing may be the most immediate way to respond to a patient's stories, successful treatments depend on a *shared* process in exploring and developing the patient's self-narrative. Retrieval by index requires far less of an effort and makes fewer demands on the therapist's attention, but it seldom gives access to the patient's experiences. For this to occur, the therapist must have the capacity to attend to the patient's disclosures; this precludes many habitual, everyday responses.

Clinical considerations

After several years of treatment, Jane, the patient quoted at the beginning of this chapter, has entered a new phase of expressing her sense of self. She is still with her husband, but she no longer speaks of her marriage in the same way. In fact, her story about herself now begins as the third child in a working-class neighbourhood, with a mother who hid from the world that she was adopted and a father with old-fashioned Protestant principles he did not always follow. Her story, still unfolding, includes many mental travels and rediscoveries of previously unavailable memories; she now accesses a central narrative with ease, without having to know the details of the therapist's personal life.

Although Jane's stories initially were well-rehearsed narratives reflecting how she functioned as a child, it took some time for the *moment of meeting* to occur and for her self-narrative to begin to form. This initial phase of her treatment could be understood as the reliving of a childhood triangular drama to win the father's exclusive attention. It could also be understood as a repetition with the therapist of a childhood pattern of trying to engage a distant, unavailable mother. In the end, it was Jane, not the therapist, who was cast in the role of storyteller, and her version may or may not include these details. What she needed the therapist to provide was the relational environment in which she could gradually articulate her own story.

For a treatment to be successful, patients like Jane need access to autonoetic awareness and the ability to reflect on highly subjective memories. Over the duration of the treatment, a self-narrative develops, told in first person and describing real events and how they were experienced. The telling of this self-narrative does not occur instantaneously, but in instalments. The process takes time and patience and occurs in the form of numerous mini-stories. However, the treatment, when successful, will be held together by an overarching and consistent narrative that creates integration. Such self-narratives have distinct storylines suited to the patient's developmental history, culture, and personality. In many respects, they are what all psychotherapy with a successful outcome has in common.

Clinical experience certainly appears to confirm narratives and memory as closely interrelated. Most uncovering of traumatic experiences relies on the patient's having some form of narrative understanding, even if that understanding is initially void of emotions. Integration and healing into a full autobiographical memory appears to require it (Siegel, 1999, 2003).

These formulations about story-based memory could be further tested by psychotherapy research, which, in fact, is well equipped to explore how therapists' stories are retrieved from memory. Many therapists' stories are scripts seldom verbalised, but they are probably accessible via immediate probing. Since therapists are generally trained to be introspective and are aware of the temporary nature of some memory retrieval, postsession interviews would give the best opportunities to explore this type of memory. Psychotherapy researchers routinely use such follow-up interviews to explore dynamics and contents that otherwise would go unnoticed; thus, the format for such explorations is already well established (Miller, Luborsky, Barber, & Docherty, 1993).

CHAPTER FIVE

Dreams as stories

Reporting and interpreting dreams is another, yet different, form of storytelling in psychotherapy treatments. Patients who remember their experiences while dreaming are often eager to share these experiences with their therapist. And, if the therapist asks, a recent dream may surface even in the initial interview. Such initial dreams, or *herald dreams*, have a particular prognostic relevance for some therapists, for reasons yet to be established beyond the clinical experience (Kradin, 2006). Aside from their use in initial assessments, therefore, dream reports may give us important insights into how stories are constructed and how they are retained in memory and later retrieved. This, in turn, should allow us to further explore the narrative aspects of psychotherapy treatments.

Until recently, investigation of dreams was hampered by a focus on the broader function of dreaming rather than on dreams' obvious narrative character (Domhoff, 2005, 2010). Today, beyond the important neurological function of sleep and dreaming, findings also indicate that dreams are similar to everyday stories (Hobson, 1994). Although some critical questions still remain, we are now able to compare the possible neural pathways for both (Solms & Turnbull, 2002). For instance, findings from research on cognitive image schemas signify that these schemas play a central role in how dreams are constructed. Since they develop very early in the infant's life and before the acquisition of language, they are often reflected in the metaphors common in dream reports (Mandler, 2004, 2005).

Although there is no simple way to tell how a dream—as experienced during sleep—differs from the actual telling of it, reports of dreams have many structural elements in common with stories in general: both have certain sequenced events and involve characters and scenes through which these characters move. In other respects, dreams are stories with some unique features. Not only are they perceived as different from everyday stories when shared with another person, they do not seem to elicit the same questions about veracity.

A critical element in developed stories is their storyline, or plot, which shapes the stories and facilitates access to the memory systems encoded with them. In dreams, more than in other narratives, these storylines are activated without input from a listener. As I will show, the storylines of dreams, in fact, appear to function independently of memory material and independently of language. In this respect, storylines are easier to identify in dreams, which by and large communicate in images rather than through verbal vehicles.

Paradoxically, not only are dreams stories we create, they are also stories created for us. And when they are used in a treatment, they become yet another narrative account of the process, thus offering both therapist and patient a new perspective on what they are exploring. *I am therefore proposing that patients' dream reports, through their metaphoric and symbolic expression, are effective shorthand communications about these patients that the therapist can use to reference their already established self-narratives.*

The nature of plots

The fact that dreams have story formats was noticed by C. G. Jung when proposing that "there are a great many 'average' dreams in which a definite structure can be perceived, not unlike that of a drama" (1948b, p. 294). First, according to Jung, dreams seem to have an exposition consisting of a statement of place and a statement of protagonists. This is followed by the development of the plot, in which tension, confusion, or conflict becomes evident. This phase, however, is followed by a culmination, or turning point, in which some decisive action or event changes the situation. The last phase, finally, offers a solution or result.

This narrative schema was first articulated by Aristotle (1971) when discussing drama and epic poetry. Contemporary scholars often regard his writings on drama as describing what they call *narrative competence*, something similar to the linguistic competence we associate with speakers of a language (Culler, 1997). As we saw in the previous chapter, they also postulate that such competence depends on being able to appreciate the structural elements in a narrative, especially its storyline.

To Jonathan Culler, a scholar of narratives, plots have features similar to those Jung ascribes to dreams. Even in translation from one language to another, the meaning of a short story, a comic strip, or a silent film will remain the same. Culler writes:

> writers and readers shape events into a plot in their attempt to make sense of things. From another angle, plot is what gets shaped by narratives, as they present the same "story" in different ways. So a sequence of acts by three characters can be shaped (by writer and readers) into the elementary plot of heterosexual love, where a young man seeks to wed a young woman, their desire is resisted by parental opposition, but some twist of events allows the young lovers to come together. This plot with three characters can be presented in narrative from the point of view of the suffering heroine, or the angry father, or the young man, or an external observer puzzled by these events, or an omniscient narrator who can describe each character's innermost feelings or who takes a knowing distance from the goings-on.
>
> (1997, p. 86)

What all versions of the plot have in common is that a transformation must occur. The plot spells out how this will happen: there is an initial condition, possibly a conflict, which leads up to some type of reversal that imposes a change, and in the end a resolution will manifest that makes the change significant (p. 85).

These elements are all present in the dream, below, of a thirty-two-year-old single male entering treatment. He told the following dream after an initial consultation:

> *It takes place in Paris. I don't know why that is, but in the dream I am discovering a highly elevated part of the city where old men in long beards are*

meeting for what appears to be a funeral. At the same time, they celebrate the return of a famous French folksinger from captivity by Islamic terrorists. They treat him as a hero and he brings with him several children born during his captivity. He has learned Arabic and can communicate with people in the Arab world. I am climbing this hill and must hold onto the side rails in order to do so.

What is striking about the dream is that only at the end do we encounter the dreamt self, the first-person self of the dreamer. Most of the dream is about characters other than the dreamer. Nevertheless, we find a statement of place in the first two sentences ("It takes place in Paris. I don't know why that is, but in the dream I am discovering a highly elevated part of the city where old men in long beards are meeting for what appears to be a funeral"). This is followed by the development phase ("they celebrate the return of a famous French folksinger from captivity by Islamic terrorists") and the turning point ("They treat him as a hero and he brings with him several children born during his captivity"). The solution, which Aristotle calls *lysis*, is contained in the concluding statement that the singer "has learned Arabic and can communicate with people in the Arab world". The last sequence introduces the dreamer's self, but again uses the symbol of the hill or an elevated part of Paris, thus tying the last sequence to the main part of the dream.

The telling of dreams

We have no reason to assume that the patient made any intentional changes in how he experienced the dream.[1] When asked to comment on how it felt waking up from it, he said the pleasant feeling had surprised him. This contrasted sharply with his depressed mood for most of the session, in which he discussed his experiences in college. (Due to immune system failures, he had had to discontinue his studies and he still felt devastated by how he had been treated.) After telling the dream, his mood became lighter and he was curious as to why he was transported to Paris and what the old bearded men at a funeral had to do with his current life situation. Could the dream be revealing certain possibilities the patient is unable to grasp, as suggested in Jung's approach to initial dreams (1934a)? His mood after telling the dream certainly differed from his mood while he described his current life circumstances. As to the *lysis*, it introduces the dreamer himself for the first time, and he now has to climb a hill with great effort. Is the dream perhaps forecasting that the treatment he is entering will require a great deal of effort on his part and steady support by the therapist?

A mood change such as my patient experienced is often the result of telling a dream for the first time. Telling one's dreams is a particular form of narration in that the teller's focus is exclusively on recounting something that happened to him or her alone and no other witnesses (Hall, 1953). Telling a dream for the first time is also an act of creating a new story; and, as described in the previous chapter, creating a story takes competence, effort, and motivation. Upon completion of this act, the teller often feels gratified, as well as surprised, by the result. Not only was a unique story created, but the teller also travelled mentally; and, in accessing memory, different feelings emerged.

In order to reenter the dream, the teller must be willing to claim that he or she witnessed the unfolding of the story; in fact, *was* the dreamer. Although the events portrayed may have little or no direct correspondence to events in the life of the listener, the teller must feel convinced of the story's authenticity. As with the telling of any new story, dream telling is, in itself, an event worth remembering.

In the process of telling about the dream experience, the teller will most likely become aware of imagery that at times is fantastic and embarrassing, even repulsive. Thus, the normalisation of the experience as "just a dream" is perhaps another reason to tell it. In this way, the experience can be placed in the past—and, importantly, categorised as something quite apart from everyday reality.

Unfolding, effortless stories

As in all storytelling, the storyline is what brings the many separate images together; this storyline is usually quite clear, even if the accompanying emotions are intense. David Foulkes, who studied children's dreams for over forty years, puts it this way:

> Momentarily, there is coherent imagery, rather than some kind of phantasmagorical jumble. Sequentially, there is a story or narrative that generally carries characters, setting, and events along a coherent course across time.

(1999, p. 126)

As with all narration, telling a dream is a recounting of something that has already happened; but where the dream narrative differs from other narratives is in how the original experiences occurred. *The dream itself happens without effort, or voluntary input, and thus unfolds without the dreamer's control* (Foulkes, 1999). The only control is to become awake. In fact, the absence of effort and voluntary input has certain things in common with being intensively involved in watching a theatre performance or a movie—narratives we become engaged with by choice (Kradin, 2006). In the latter instances, the control offered us is to cut short any full mental or emotional participation and let critical faculties resurface; we also have the option of leaving the theatre or turning off the DVD player or computer. However, when we are fully engaged in viewing a performance, we do so—similarly to when we are dreaming—as engaged but passive recipients with no power to manipulate the narrative.

In spite of our lack of control over content and outcome while we are dreaming, when telling the dream we experience a distinct narrative, with a beginning, a middle, and an end. Even though we may possibly change some scenes and forget certain parts, telling the dream is similar to telling about a movie we watched in waking life. This may explain why a dream is sometimes mistakenly called "the movie". As in watching a movie, the dreamer is clearly cast in the role of spectator—although, upon awakening, when giving the dream language, the dreamer is also the dream's creator.

The left brain: "the interpreter"

The one part of dreams that appears to have been located by neuroscientific research is the storyline. According to Michael Gazzaniga of Dartmouth College and the University of California at Santa Barbara, strong experimental evidence now shows that the left brain "is crammed with devices that give humans an edge in the animal kingdom" (1998, p. 132). He calls these devises "the glue that unifies our story and creates a sense of being a whole, rational agent" (p. 152).

Gazzaniga's research, conducted with his colleague Joseph LeDoux, is based on studying split-brain patients in whom the connection between the left and right sides of the brain is severed.[2] Surprisingly, the neural networks in the left frontal lobes of these patients continue to provide elaborate but spurious information from whatever data they can find in other parts of the brain, even when critical input from the right side is no longer available. Gazzaniga writes:

> Relying on the experimental approach I describe in Chapter One where we showed a picture of a chicken to the left brain and a snow scene to the right, we set to work and discovered that the left brain has a specialized mechanism that interprets actions and feelings generated by systems located throughout the brain.
>
> (1998, p. 133)

Gazzaniga discovered this mechanism, which he calls *the interpreter* and which appears to function independently of any input from our perceptions of outer environment and social reality—thus, in a manner similar to narrative plots.

> The interpreter constantly establishes a running narrative of our actions, emotions, thoughts and dreams. It is the glue that unifies our story and creates a sense of being a whole, rational agent. It brings to our bag of individual instincts the illusion that we are something other than what we are. It builds our theories about our own life, and these narratives of our past behavior pervade our awareness.
>
> (Gazzaniga, 1998, p. 174)

As a continuously running narrator, the interpreter seems to have access to whatever storylines a person has at his or her disposal. Even when there is no real input from external experiences, as in dreaming, the interpreter continues to operate. Although certain conditions are different from those of split-brain patients, when a person is dreaming cortical activation persists, involving both recent episodic memory traces that have not yet been consolidated and more remote episodic traces. This activation, moreover, appears to be independent of the retrieval cues necessary when we are awake. The interpreter, with its storyline explanations, has plenty to work with, even when there is no perceptual input.

Paradoxically, the interpreter has little to do with another important function located in the left side of the brain: articulation of language. In their experiments with split-brain patients, Gazzaniga and LeDoux relied exclusively on visual images. When they showed the pictures of a chicken and a snow scene, no language was involved. Thus, their research confirms the observation by Culler and others studying literature: that storylines are by and large independent of language (Culler, 1997).

The independence of the left-brain interpreter may also explain one major difference between the dream narrative and narratives produced when one is fully awake. The bilateral cooperation missing in the split-brain patients—and the lack of perceptual input when one is dreaming—precludes what Daniel Siegel identifies as *full neural integration*. For this to occur, both left and right hemispheres must participate. Siegel states:

> The left hemisphere's drive to understand cause-effect relationships is a primary motivation of the narrative process. Coherent narratives, however, require participation of both the interpreting left hemisphere and the mentalizing right hemisphere. Coherent narratives are created through inter-hemispheric integration.
>
> (1999, p. 331)

What the interpreter thus seems to produce is a certain type of logic that is syllogistic and applies to most narratives. It is the logic of story, of conceiving how one thing leads to another, and of "how something may have come about" (Culler, 1997, p. 84). In Gazzaniga's account, it is also "the capacity to state a major premise, followed by a deductive conclusion", which "is what our species alone can do" (1998, p. 152).

The search for function

The fact that dreams appear to be a mostly involuntary part of sleep has led to a great deal of speculation about their function in our overall mental life. Freud, in his groundbreaking *Interpretation of Dreams*, assumes that dreams protect the dreamer's sleep (1990a). For Jung, dreams compensate for inadequacies in the dreamer's adaptation to current reality (1948b). Both proposals have been questioned by more current data, and other proposals have proven equally elusive (Christos, 2003; Lavie, 1996).[3] The most recent proposal—that dreams serve to place certain experiences in long-term memory and that REM sleep, in which most dreams take place, improves memory storage and retrieval—still lacks proof (Lavie, 1996; Rock, 2004).[4] Allan Hobson of Harvard University, although prefacing his discussion as "still speculation", gives a modified version of this theory:

> We sleep, and the past day's memories are reactivated as we dream, which changes their status; it advances them from short-term memory into long-term memory, perhaps by imposition of acetylcholine, which is omnipresent during sleep.
>
> (1999, p. 115)

This and other physiological explanations of dreaming are still incomplete and often have serious speculative shortcomings. Hobson's theory (1988) that dreaming is a form of random visual and auditory hallucination renewed interest in physiological explanations of dreaming. After all, Hobson and his colleague, Robert McCarley, had earlier demonstrated that dreaming originates in the brainstem (as activation of cholinergic mechanisms in the pontine brainstem) and therefore may have nothing to do with higher cortical functions (Hobson & McCarley, 1977). Several other researchers have since proven this conclusion to be incomplete (Revonsuo, 2003); Hobson has also modified his earlier conclusions from 1977. Reviewing the consequences of lesions to the brainstem, for instance, Mark Solms, another dream researcher, found no indication that these lesions affect dreaming, though lesions in the frontal lobes do (1997). Dreaming, although originating in areas controlling sleep, eye movement, heart rate, and breathing, must have wide-ranging activation reaching into other areas as well. And dream work itself seems to occur in higher cortical brain sites.[5]

In spite of these obstacles to establishing the full neuroanatomy of dreaming, certain facts are known today that neither Freud nor Jung had access to. Serious research on sleep and dreaming began with the 1953 discovery of REM sleep by Eugene Aserinsky and Nathaniel Kleitman of the University of Chicago. Based on current data, William Domhoff of the University of California at Santa Cruz, who specialises in the cataloguing of themes in dreams, argues that dreaming and waking cognition have more similarities than differences, and that waking fantasy, daydreaming, and dreaming have strong connections among them. He writes:

> Dreams are a dramatic and perceptible embodiment of schemas, scripts, and general knowledge. They are like plays that the mind stages for itself when it doesn't have anything specific to do. In

particular, many dream scenarios express several key aspects of people's conceptual systems, especially self-conceptions, which can be defined as a set of cognitive generalizations about the person that guide the processing of self-relevant information and events.

(2010, pp. 1–2)

Some dreams of a shorter variety may no doubt fit the description of a *schema*, or *script* (see Chapter Four). However, in what Domhoff calls their "perceptible dramatic embodiment", dreams are most accurately characterised as narratives, or stories. First of all, they usually adhere to the plot format described above. Frequent references to episodic memory traces occur, especially in fully developed dreams, in which self-conception plays the larger role. Thus, they have much in common with the self-narratives described in the previous chapter. Finally, in referencing a dreamer's self-conception or *working self*, they are auto-biographical. By all accounts, dreams are, indeed, "like plays that the mind stages for itself", as Domhoff puts it.

The content of dreams

These common characteristics do not mean that the content of dreams is predictable. What is selected from our past experiences may not always be apparent. When the current concerns of female college students, gleaned from pretest interviews, were compared to the actual reported dreams that the participants had the night following the interviews, it was found that the researchers were not able to predict what the participants were going to dream. They were, however, able to identify the topics of the dreams as related to common themes in the students' lives, such as academics, family, and friends (Foulkes, 1999; Roussy, Camirand, Foulkes, De Koninch, Loftis, & Kerr, 1996).

From this point of view, what is significant about dream stories is that *they are stories that were never told before and stories that are told without any apparent input from the dreamer's wakeful consciousness.* In dreams, we are able to observe the creation of a new narrative about the person's self, often with unexpected and surprising details, turns, and characters. Dreaming, in the words of Foulkes, "creates patterns that have never been experienced before and it does so in the absence of environmental stimulation" (1999, p. 11).

What is also significant, especially considering the experience of dreaming, is that dreaming mainly occurs in images (Hall, 1953). Only in one's telling of a dream does language enter into the picture in any significant way. This must imply that our conceptual knowledge plays a less important role, if any at all, in the production of a dream. However, as Hobson points out in his discussion of one person's dream (Delia), "her sense of being in the scene and *moving through it*" shows the dream as both constant and vivid (emphasis in original). The dream narrates, through motion, from one distinctly arranged scene to the next (1999, p. 39). While a sense of speech and touch only enters more sparingly, both vision and motion are continuous.

The mechanics of dreaming

After years of research, the mechanisms responsible for dreaming now seem established (Foulkes, 1999). Dreams are what happens when core aspects of consciousness are still in operation, while other aspects appear uninvolved or simply shut down. The following three conditions seem to be present:

1. persisting cortical (physiological) and cognitive (psychological) activation
2. occlusion, or dampening, of external stimulation
3. relinquished voluntary self-control of ideation.

The first of these conditions, cortical activation, tells us that some kind of conscious ideation is experienced when one is dreaming. The second condition means that during dreaming the dreamer does not process perceptions of the outer world. Not only are the dreamer's eyes closed, the sleep mechanism in his or her brain shuts down the processing of all external stimulation impinging on it (Hobson, 1999). And, as the result of the third condition, we loose voluntary control of our ideation. Foulkes summarises:

> As any insomniac can attest, failure to be able to let go of self-awareness and self-control prevents sleep from happening. And so, one price of achieving sleep is losing the self that voluntarily, intentionally, regulates our waking mind. It is not just the world

that helps us organize our waking thought, but also our conscious intentions about how we want to address the world.

(1999, p. 125)

This means that there is no longer an everyday self (or *ego consciousness*, using the older terminology) at work, suppressing what seems irrelevant and directing attention to what seems most relevant. Unexpected memories will appear, while more obvious ones will not. Long-forgotten events show up side-by-side with very recent ones, sometimes even merging the faces of two people known at different times in the past, in a process called *transmogrification* (Hobson, 1999).[6] It is as if the dream wishes to synthesise episodic memory traces from across a lifetime and it does so without everyday consciousness directing it.

When we are listening to someone telling a dream, this poses a dilemma about what standards to apply to what we hear. Listening to someone tell about recent experiences in the form of a story, we tend to focus on veracity. We ask ourselves if the story seems to hold true or if the person is inventing details. In contrast, when someone volunteers a recent dream, these standards no longer apply. Instead, we may wonder about the sanity and well-being of the dreamer, especially if some details appear bizarre and involve sudden changes in time, characters, and setting. We may look for a way to establish that, in spite of the dream's peculiarities, the person appears well grounded in reality. The truthfulness of the events in the dream is not our first concern, and we would be unlikely to apply other criteria common to stories; questions as to what, in reality, may have occurred are irrelevant.

Lesion studies and dream pathology

One of the few ways to explore the specific neuroanatomy of dreaming has been to study patients with brain lesions and how these lesions affect dreaming. In his studies of 361 patients with various lesions, Mark Solms (1997), a South African researcher also trained as a psychotherapist, describes four common syndromes in which dreaming was disturbed or distorted due to brain damage:

- *global anoneira,* or total loss of dreaming
- *visual anoneira,* or restrictions on (or absence of) visual imagery

- *anoneirgnosis,* or confusion between dreaming and reality
- *recurring nightmares.*

One of Solms's findings is that the parietal lobe, located in the forebrain, plays a major role in dreaming. This region combines different forms of sensory information and creates a sense of spatial orientation as well as mental imagery; it thus appears critical to the experience of dreaming. When this region was damaged, these patients experienced no visual imagery, in spite of activation of dreaming in their brainstems.

Dreaming also involves both brain hemispheres; damage to one cortical structure often leads to loss of visual imagery. Patients with damage to both sides of the prefrontal cortex ceased dreaming altogether. However, disorders of spoken language do not, by themselves, affect dreaming; thus, dreaming is not dependent of those brain regions responsible for the ordinary production of language. Most significantly, perhaps, Solms was also able to show that dreaming involves what appears to be an autonomous system of vision. Cortical areas of the parietal lobe that are necessary for processing external visual information do not appear involved in the production of normal dream imagery.[7]

Solms (1997) also concluded that the inferior parietal structures are involved in spacial cognition in dreams. Any damage to this region, to either the left side of the brain or the right, stops the subjective experience of dreaming. Thus, spacial cognition must be fundamental to dreaming. Generated by heteromodal syntheses in the parieto-occipito-temporal junction, right parietal lesions appear to affect concrete spatial cognition, while the left side appears to be involved in symbolic (quasi-spatial) cognition, although the former appear to play a more essential role.

These findings have, by and large, been confirmed by more recent brain imaging (Braun et al., 1997). Of particular interest is also Solms's (2000) theorising about the visual imagery in dreams which he bases on computational modelling by Michael Kosslyn of Harvard University. According to Kosslyn (1994), dream images, as well as other images, "are fed backwards into the cortex as if they were coming from the outside" (1994, p. 75).

In Solms's (1997) research, the following areas are inessential, or inhibited during the dreaming process: the spinal and peripheral sensory-motor systems, the primary (idiotypic) sensory-motor cortex, the unimodal (homotypical) isocortex outside of the visual sphere, and the dorsolateral prefrontal cortex. Other brain regions seem to have only minor roles, by activation or inhibition.

When translating these findings into how dreams function in a clinical setting, the following conclusions can be drawn:

1. Dreaming involves activation of both hemispheres and loss of dreaming may result from lesions in either hemisphere.
2. Complete cessation of dreaming (*global anoneira*) may occur from lesions in several regions of the brain; thus, dreaming is modular and occurs in several different regions (Solms & Turnbull, 2002).
3. Dreaming appears independent of ordinary language; patients suffering from aphasia and loss of spoken language still appear to dream.
4. Dreaming involves a visionary system that is independent of eyesight. Thus, those brain sites in the parietal lobe responsible for processing external visual stimulation are not involved in normal dreaming.

As a result of his lesion studies, Solms also concluded that dreaming is not synonymous with REM states, a conclusion confirmed by functional brain imaging studies (Braun et al., 1998).[8]

A philosophical dilemma

In many ways, the different conclusions about physiology and functioning of dreams reflect a philosophical dilemma that goes back to the pioneers of psychoanalysis, or possibly even further.[9] Depending on a person's tolerance for ambiguity and fantasy, for imagination rather than rationality, two clearly different views exist concerning dreams—both claiming to be based on evidence, both advocated with great conviction that must reflect commonsense biases or beliefs.

Freud, for his part, assumed that the dreams "replace thoughts by hallucinations" (1900a, p. 50). Thus, at least in his basic understanding, he equates the dream's visual images with psychotic symptoms since images appear to disguise actual thoughts or wishes:

> The transformation of ideas into hallucinations is not the only respect in which dreams differ from corresponding thoughts in waking life. Dreams creates construct a *situation* out of these

images; they represent an event which is actually happening; as Spitta (1882, p. 145) puts it, they "dramatize" an idea.

(1900a, p. 50; emphasis in original)

To Freud (1916–1917), the dreamer believes that he or she is experiencing the otherwise-forbidden thoughts that the dream has made into images. This is what he calls the *dream work*, "the work that transforms the latent dream into the manifest one" (p. 170). However, this process can also be reversed by the therapist's interpretations.

> The work to proceed in the contrary direction, which endeavors to arrive at the latent dream from the manifest one, is our work of interpretation. This work of interpretation seeks to undo the dream work.

(p. 170)

The meaning of the latent dream is, therefore, something that is known to the therapist as an infantile wish unacceptable to the full awareness of the dreamer. From a therapeutic point of view, there is thus nothing new to learn from the dream that the therapist is not assumed already to know. At least in terms of Freud's metapsychology, infantile wishes of an incestuous nature are the cause for most psychological disorders (Freud, 1916–1917).

Unfortunately, in Freud's understanding of dreams there is no acknowledgement of the fact that, by nature, they are narrative, not merely dramatised ideas. The therapist, as the listener, holds all the cards as the only participant who knows the secret when the dream is told: all dreams express an otherwise-unacceptable wish (Freud, 1917d).

Jung, on the other hand, had a more intuitive understanding of dreaming. He regarded dreams as "highly objective, natural products of the psyche, from which we might expect indications, or at least hints, about certain basic trends in the psychic process" (1935b, p. 131). Thus, for him, dreams arise spontaneously "without our assistance and are representative of a psychic activity withdrawn from our arbitrary will" (p. 131). By amplifying the images in a dream to other narratives, especially those in tales and myths—which, presumably, are generic expressions of narrative structures—Jung was certain that the dreamer could gain insights into his or her own psychology. Jung also claimed that dreams could provide unique perspectives since they reflected certain patterns or archetypes, inherited tendencies of an evolutionary nature. Many dream images could therefore be regarded as based on such archetypes.

Jung gradually abandoned the traditional psychological distinction between a dream's manifest and latent content. Dream symbols became associated with archetypes as inborn patterns of meaning (1935c).[10] However, like Freud, he continued to speculate about the function of dreams and assumed that dreams were prospective.[11]

> The occurrence of prospective dreams cannot be denied. It would be wrong to call them prophetic, because at bottom they are no more prophetic than a medical diagnosis or a weather forecast. They are merely an anticipatory combination of probabilities which may coincide with the actual behavior of things but need not necessarily agree in every detail.
>
> (Jung, 1948a, p. 255)

This understanding seems to have its origin in Jung's early interest in initial dreams (1913a, 1913b). In his description of them as having some highly unusual properties, "often amazingly lucid and clear-cut", he indicated that these dreams carried a special significance, while those reported well into the treatment could "lose their clarity" due to transference/countertransference dynamics (1934a, p. 145).

The desire to make predictions based on dreams may at times have made Jung lose sight of the obvious: their content is based on memory traces referencing the dreamer's past, whether recent or distant. Even when dreams create a surprising narrative about these traces, they do so by activating already encoded material. In this sense, their stories reflect restructuring and reshaping of stored engrams into new combinations. This in itself does not make dreams prospective or prophetic, any more than a meteorological report is a forecast based on already available information.[12]

The dreaming self

Today, these two traditions still compete, even though the authors advocating for each may, at the same time, be critical of Freud or Jung, or both. The significant issues appear to be how to define the type of consciousness present in dreams, and to what degree dreams "speak" in a bizarre and unintelligible language.

Without trying to solve the bigger issue of human consciousness in general, we can nevertheless state that dreaming requires some type of consciousness, even if this consciousness is quite fragile and often lost upon awaking (Bulkeley, 2005). This implies that there is a particular form of consciousness that reflects what I will call a *dreaming self*, comparable to a *wakeful self* as this dreaming self creates daytime stories out of a multitude of experiences. At the same time, dreaming activates an *observing self* that functions the way a listener to a story or a spectator to a theatrical performance functions.

In the dream, however, these selves operate within the same person. There is no social and relational context, no dialogue with a real "other". The dream, much like a fantasy or daydream, contains both aspects of self. In this sense, the dream reflects a certain creative process, but without an opportunity for inner dialogue. Its author, the dreaming self, is not able to modulate the flow of images and thoughts the way an author does when writing a novel—or a visual artist, when painting from inner visions. Dreaming appears to require that the two selves function in prescribed roles on a permanent basis.

In most dreams, the observing self is also able to identify with the *dreamt self*; the result is that we feel that we are dreaming about ourselves. This, in turn, makes the dreamt self in the dream narratives function the way the self in self-narratives functions, or as the working self functions in autobiographical memory (see also Chapter Four). We dream about ourselves the same way we tell stories about ourselves and our experiences. The difference between the two is that we tell stories about ourselves to others so that we will remember what happened to us; whereas, in dreams, a story is being told *to* us and this story, for all its symbolic bizarreness, is harder to remember the same way. It has to be told to us before we can retell it in words; and unless we retell it by translating it into a language we can remember, we will forget the dream experience.

Dream memory

In spite of our difficulty in remembering what we dream, our ability to travel mentally in time remains intact. This would point, therefore, to our access to memory for past events while we dream. Even though many of the things we associate with waking consciousness are missing, such as control over ideation and the use of language and concepts, our episodic memories seem to be activated, even if in a highly disorganised fashion (Foulkes, 1999).

The primary sources of information still available to the brain while we are sleeping are, therefore, past experiences and the current state of the organism. From this internal information, dreams create what Stephen LaBerge of Stanford University calls "a simulation of the world in a manner directly parallel to the process of waking perception minus sensory input" (1998, p. 495). Dreaming is the result of the same perceptual and mental processes that we use to comprehend the world while awake.

Two features in particular point to episodic memory as one of the functions still in operation when we are dreaming. Most remembered dreams are told in the first person. The dreamt self has the capacity to travel in time, and thus seems to possess autonoetic awareness. However, we appear to have only temporary access to the material that surfaces when we dream; if we do not memorise our dream shortly after having it, the entire experience will usually be lost. We, therefore, have to assume that memories appearing in dreams are accessed differently than when we are in an awakened state and that the cues are accessed differently in the two states.

Our mind, when put to work in REM sleep, faces the task of, as faithfully as possible, giving representation to memory traces that may never before have been connected. Although dreaming is continuous with our waking, reflective ability to think in images and travel mentally, our perceptual competence no longer seems necessary, or is put out of work. To help in the task of creating a story from often disparate memory traces, the dreaming self has at its disposition the storylines that we have used in memorising events in waking life. In fact, when we compare what is known about narratives and dreams, the common storylines are striking. As we saw in Chapter Three, narratives are memory devices; and we have to assume that the same holds true for their role in dreams.

William Domhoff mentions three factors relevant to remembering a dream: dramatic intensity, recency, and length. In analysing these three variables, he concludes: "It was found that recency can compensate for low intensity. If these results can be generalized outside the laboratory, they suggest that recency and length are giving us many dreams not usually high in dramatic intensity" (1996, p. 43). The recency factor, in other words, is closely related to episodic memories, the window for encoding them, and the context for their retrieval, while length appears to be closely related to well-developed storylines (see further, Chapter Four).

The storylines in dreams are more fixed than in regular narratives. If, for instance, I wish to tell a friend about a recent visit to California, I may include certain details because I know he is interested in wine and the trip included visits to vineyards in the Napa and Sonoma valleys. I may also, as his interest is being peeked, tell about certain wines I tasted. However, when I tell a colleague about the same trip, I may include other details, like a visit to the Berkeley University art museum, since I know that the colleague has a particular interest in modern art. I may even change or invent new beginnings, endings, and other aspects of the plot.

Something quite different happens when I wish to tell someone a dream. To begin with, I am much less likely to plan the disclosure. Only when reminded of some portion of a dream, usually by association with the topic of conversation, might I reveal certain of its details. And if, in a more unusual circumstance, I tell a whole dream, I have very few options when it comes to editorialising and fitting it into what I think the listener may be interested in, unless I've already told the dream. Once I begin telling the dream for the first time, my attention is fully occupied by the effort to remember the images and to verbalise them. The cues are primarily the ones built into the dream narrative rather than from the feedback I perceive from my listener, and the narrative is less open to change. As an experience, the dream is usually not a conversational item in itself, but belongs to a different form of consciousness in that it is memorised in a prearranged but less durable way. The story being spun is an inner dialogue of a rather involuntary nature. It is a story between me as the dreamer and myself as the observing participant.

Dreaming and development

One of the surprising discoveries in David Foulkes's extensive research on children's dreams is that dreaming may not be present at birth, at least not in what can be recorded using electroencephalography (EEG) and identifying REM periods of sleep (Foulkes, 1999; Rock, 2004). Before a critical ninth birthday, children also appear to lack dreams similar to those of adults. The material they are able to recover before age four to five is typically cryptic and without much of a storyline (Foulkes, 1999). The following examples were recorded upon the participants' being awakened in Foulkes's dream lab:

> (Dean, at age four years eight months): I was asleep in the bathtub. (Several months later): I was sleeping at a co-co stand, where you get Coke from. (At age six years eight months): A cabin at Barbara Lake. It was little, and I looked in it. Freddie [friend and fellow participant in the study] and I were playing around, with a few toys and things. (One month later): Some guy swimming, and baseball at high school. (Two days later): Cows on the ranch, just mooing, just running around. I could see the ranch house.

> (pp. 159–160)

The first two dreams, as well as the last two, are scriptlike; and the first two seem to refer to the research situation itself. There is no apparent motion before the third dream's "looking in" and "playing around", and no speech is mentioned. Only in the forth dream is there any mention of sound. In all five dreams, the dreamt self ("I") is mostly passive or nonexistent.

Between ages seven and nine years, the dream reports become more complex and certain plot structures now seem to be part of the dreaming. However, an active and reflective dreamt self seems to emerge only after that age. Foulkes gives the following examples from the same boy:

> (Age eight years eight months): We were supposed to come here and sleep. Only, when we came here, we had a party and Freddie [same time, but different friend from above] got to bike. Freddie, and John, and Kate, and Freddie [the first one] were there. You were in your office, putting on the wires. They had a big cake, and everybody was giving out gifts. I got a weight-lift. (Two months later): One day, my mom, she got my sister some skis—she got all three sisters some skis—and I didn't get any skis. It took too long, and they didn't have any small enough sizes. And she got a letter, that we could go skiing up at Steamboat Springs, and I couldn't

go, but when they got back, they had gotten me some skis. Then I got to go skiing up at Happy Jack [a closer and less grand ski area] with them.

(pp. 160–163)

In this set of dreams, reference is again made to the dream experiment itself. In the second dream, about being given skis, for the first time there is a storyline or plot that presents a problem and allows for a turning point in the story as well as a resolution.

The findings from Foulkes's research confirm findings in research on children's development. Robyn Fivush of Emory University concludes that, before a certain age, the child relies on scripts representing "what happens each and every time the event occurs" (1997, p. 142). Such scripts are not abstractions of an explicit and semantic nature and do not give access to memory of specific episodes in the past; thus they would preclude memory of dreams, possibly also dreaming, in clearly narrative form (Foulkes, 1999).

From age four to eight, however, children acquire the ability to form autobiographical narratives—a specific, long-lasting, and particular achievement fostered by social conditions. The capacity to form episodic memories, a necessary component in autobiographical memory, has a similar late arrival (Siegel, 2003). As Kathrine Nelson and Robyn Fivush (2000) show, the child's representational system is, at this point, accessible to linguistic formulations presented by other people. A sense of having a unique and personal history is achieved. This is made possible by the development of a narrative competence that allows children to form lasting, explicit memories. In this respect, autobiographical memory is tied to language. It has the dual function of being a mental representation as well as serving a need to communicate (Nelson, 2000). It is also tied to the capacity to form episodic memories, which in turn makes fully developed dreaming possible.

Nightly hallucinations

Since dreaming occurs in images, many observers have found dreams to "speak" a bizarre and unintelligible language. Terms like *hallucinations*, *delirium*, and *delusions* have been used in proposing that the formal elements in dreaming are accidental. As we saw above, both Freud and Hobson refer to the imagery in dreams as "hallucinations". Hobson titled one of his books on dreams—in which he compares certain findings about the brain chemistry in psychosis with the brain chemistry during dreaming—*Dreaming as Delirium* (1999).[13] Solms and Turnbull (2002), in defending Freud's notions of dreaming, try to justify the idea that the content of dreams, as they are reported in dream research, appears to be both bizarre and unintelligible unless we assume a deeper, latent meaning.

Both these conclusions have been rejected, based on extensive analyses of dream reports. Referring to a series of large investigations—over a period of seven years—conducted in two different laboratories in the 1960s, Domhoff (2010) concludes that as many as ninety per cent of the reports could be considered descriptions of everyday experiences. Sixty to eighty per cent of these reports as were rated "high in coherence"; sixty-five per cent, "highly credible"; and fifty per cent of the long dream reports were rated as "having no bizarreness". Only ten per cent were rated "high in dramatic quality"; and only in two per cent of the material did the raters find a high degree of bizarreness.[14]

The notion that dreams are too bizarre to be understood also contradicts clinical experience. Patients in analysis vary significantly in their ability to recall and appreciate their dreams. As Paul Lipmann, a seasoned psychotherapist, notices in his book *Nocturnes*, there are "dream adorers" and "dream despisers": those "who listen intently to their dreams, measure their messages, weigh their consequences, record their minutest activities", and those who, "deaf to their dreams, care little about the goings-on unless shaken by a powerful dream thrust" (2000, p. 84). According to his observations, female patients in general care more about their dreams than male patients; and cultural traditions as well as personality traits seem to influence the ability to report dreams.

From my observation, patients also differ in their ability to associate with the images in their dreams. This appears to have little to do with the seriousness of their psychological problems; however, a significant factor is the patient's comfort with visual expressions and with metaphoric and symbolic language. Patients with regular exposure to artistic and metaphoric expressions, and those who come from a family environment in which they were an active part of communication, seem more inclined to speak about their dreaming experiences. Patients whose orientation is more rational and critical tend to be more hesitant, not only about exploring their dreams but about exploring the imagery in visual art, fictional writing, even in everyday colloquial language. This, in turn, may explain why, for certain people, dream narratives seem quite natural, while for others they are irrelevant—an attitude that may lead to less of a capacity to dream (Foulkes, 1999).

One reason for this discrepancy—between those who claim not to dream and those who find telling dreams worthwhile—may also be the exclusively private nature of the experience of dreaming. Since we have no way of recording the dream itself, we must rely on how the patient reports the experience after the fact.[15] While dreaming, however, the dreamer generally is not aware of this exclusivity. And, upon awaking from the dream, when this fact is realised, the dreamer still remembers, often in vivid terms, certain references to known persons and events that occurred in the past. However, the experience is different from standard recall, in that time, location, and participants may no longer refer to the same experience. During a dream, a person remembered from a distant past may appear in a recent location and a different context. Thus, the dream appears to create its story more as a fiction writer does than as an historian; it is more like a novel than a chronicle.

By the time patients relate a dream to the therapist, they are usually quite aware that what they report is not about anything that someone else observed. Even when referencing actual past events and persons, dreams do not meet the standards of simple reality testing; most likely, the events as portrayed could never have occurred in the way the dream presents them. What is reported to the therapist is an inner experience; thus, the patient's motivation for remembering the dream will vary considerably depending on how the patient values inner experiences in general.

We are, therefore, confronted with at least three different mental states: (1) the dreamer's awareness during his or her experiencing of a dream, (2) the dreamer's awareness during his/her immediate recollection of the same dream upon awakening, and (3) the dreamer's awareness when relating his/her dream experience to the therapist.

This means that, again, consciousness of self also undergoes changes. As we noted above, in the dream as it is being dreamt, there is both an observing self and a dreamt self. Upon the dreamer's waking up, this dichotomy becomes apparent by the dreamer's taking on the observer role while still being closely connected to the dreamt self (the dream's main protagonist). However, by the time the dreamer reports his dream to the therapist, a considerable distance has developed in regard to this close connection to the dreamt self. The patient's consciousness of social and external reality has returned and resumed its primary role. For some patients, this change leads to a complete loss of memory about the dream; while, for others, the dream experience is still available, although it had to be primed by their remembering it upon waking up.

Psychopathology and dreams

The stages of remembering do not preclude signs of psychopathology in patients' dreams. My English colleague Margaret Wilkinson makes a strong case for the dreaming process's being a dissociative phenomenon, especially in cases of severe psychological trauma (2010). In such conditions, dissociation often is present in other parts of the patient's life as well. The metaphorical aspects of a remembered dream can, therefore, be understood as representing a patient's dissociative thoughts and serve as a bridge between two different domains of his or her memory. (See also the discussion in Chapter Six about metaphors.)

Mark Blechner (2001), of the William Alanson White Institute and the New York University Postgraduate Program in Psychotherapy and Psychoanalysis, questions whether patients suffering from borderline conditions and psychoses may have more rational and "undreamlike" experiences when dreaming than their general conditions would suggest. Referencing Freud's work, he wonders if there are not two types of dreams, realistic and fantastic. He also refers to another psychoanalyst, Paul Federn (1952), who arrived at similar conclusions about the dream material of patients in melancholic states. They, too, appeared to have concretistic rather than symbolic dreams (Blechner, 2001).

How much of these observations must be ascribed to these patients' ability to give verbal accounts of their dreams is an open question. We do not have information about what happens when they verbalise what they dreamt. In other words, the phenomena observed by Blechner may not reflect particular distortions in these patients' states of mind while dreaming but, rather, their inability to reflect on their dreams when they communicate with the therapist.

In itself, dreaming may certainly be viewed as the dissociation between the dreamt self and the observing self. This fact, alone, does not qualify all dreams as forms of splitting or dissociation. Rather, the tensions between the self being dreamt and a passive, observing self appear to be the result of how the dreamer verbalises what he or she experienced while dreaming. What researchers and observers of dreams call "bizarreness" may, in my judgement, be the function of how dreaming has been characterised when one views the resulting verbalised reports. This conclusion is supported by Blechner when he writes that the dream "allows the mind to think without the constraints of language" (2001, p. 25). Foulkes (1999), when reviewing the result of his research on children's dreams, comes to similar conclusions.

Dream language and narratives

We have now identified two significant characteristics that dreams and narratives share. To summarise:

1. They both depend on access to episodic memory, even though dreams appear to access memory differently than common narratives do.
2. They also share implicitly encoded storylines, or plots, that are particularly visible in longer dreams, or series of dreams, from the same time period.

Dreams are different from most narratives in their reliance on visual language. How this imagery is transformed into concepts and thoughts presents researchers with a difficult methodological problem. Since dreaming is primarily visual, the dreamer must create verbal vehicles often referencing different cognitive domains. This, in turn, prevents firsthand knowledge of dream experiences. For the purpose of research, what is knowable are only dream reports.

This also presents a problem for the dreamer. Relying on verbal vehicles to describe visual or visuomotor experiences requires the ability to find appropriate expressions of a kind that may not be used with great frequency in our everyday interactions. The difficulty can be overcome, however, if we reverse the order. For instance, when a filmmaker re-creates a dream using its visual language, we are easily able to follow the narrative, even though we may find it frustrating to give a verbal account of what we experienced. The same is true for the ability to communicate generally about visual arts, film, and drama. The images—scenes, movement, and sounds—produce associations and bring back memories, although articulating these verbally is a form of translation from one language to another.

As Allan Hobson points out, dreams offer other challenges as well. The experience of dreaming tends to be what he calls *hyperassociative*, and the categories that are linked together during the dream are so broad as to become overinclusive. This makes it difficult to trace the particular memories involved (1994). In telling a dream, the dreamer must therefore find ways to separate the various associations, or allow for a multitude of them to exist in reference to the same image. Finding language for this exercise takes training and facility with metaphors and symbolic expression.

As long as we focus exclusively on dream reports, this problem may go unnoticed; dreams are simply another form of narrative expression. For their clinical use, however, as the nature of a particular dream becomes significant, a different problem presents itself. In relating the dream to the therapist, the patient has to use verbal vehicles and translate his or her dream from a visual language to a verbal one.[16] When a patient describes his or her life situation and memories of past experiences, this is less of a problem. The material usually relates to the patient's recent past and to experiences lived through in full, wakeful consciousness. In the ensuing dialogue, both participants (therapist and patient) easily identify time, place, and protagonists. As a result, the memories formed are easily retrieved and placed in clear time frames.

The consciousness reflected in a dream is different. It can only be communicated indirectly, in past tense, and its reporting is highly dependent on conscious skills that may not be present while dreaming. The communication, thus, depends on the patient's ability to remember and to build verbal bridges to the dream's images. This problem may not be immediately apparent to the therapist, who may assume that an interpretation is what is called for, without noticing how the patient is able to talk about it. When therapists fail to appreciate this dilemma, perhaps by insisting on examining the dream as told, patients may feel left out of the discussion and unable to stay connected to the dream as a narrative about themselves. This is especially true when the therapist amplifies individual images in the dream. The process can become quite didactic and move attention away from the dreamer and his or her way of translating the dream into a meaningful story.

In other words, there is a built-in dissociation in the experience of a dream. The dreamer is also the dreamt self; yet a patient's dream report is primarily based on how his or her observing self is accessed at a later point, in a wakeful state. This is probably what Freud had in mind when he theorised that dreams have a latent content. He may, in fact—in my view, mistakenly—have wished to account for the situation as resulting from the relational context in which the dream is being told, and thus included that context as a part of the dream. In so doing, he confused the dream itself with the dynamics that may occur in the therapy relationship when the dream is being discussed. By trying to account for both in a concise theoretical model, he ended up assuming that all dreams have the same underlying narrative meaning, which they clearly do not. Each and every dream has its own particular story to tell. Their storylines differ and their memory contents are quite varied and unique.

Remembering dreams

The fact that the experience of *self* in a dream is split into a dreamt self and an observing self, into object and observer, probably explains why many dreams are difficult to remember. In terms of neuroanatomy, our ability to remember and give language to what occurred when we were dreaming seems to depend on the integration of a left-brain storyline with the right-brain processing of contextual information. The window for this recall is usually quite narrow. Unless dream experiences are captured and verbalised shortly upon awaking, they are usually lost for good. Although some strongly motivated dreamers regularly record their dreams for themselves, most dream reports come from patients in psychotherapeutic treatments in which remembering them is encouraged and serves a therapeutic function (Domhoff, 2010). Even so, much dreaming that does not involve a clear narrative about the dreamer's self will generally not be remembered.

A curious exception is when the dream is cued by a question from the therapist. As Blechner notices, such spontaneous recollection appears to show what he calls "an unconscious metaknowledge of the connection between the dream and the question" (2001, p. 244). As I will show in the next chapter, this phenomenon is consistent with what we know about cueing of episodic memory. Certain dream metaphors may override the otherwise typical memory barrier. In fact, the cues to remembering a dream, like those for other memory, are mostly context dependent. The cues may still be available when the precise circumstances and associations can be repeated, although the dream as a whole appears forgotten. This would also explain another common and spontaneous activation of dream images: when associations to a particular person who appeared in last night's dream surface during daytime consciousness.

That there is a barrier between one's consciousness while dreaming and one's consciousness while fully awake has been well documented, but this barrier is by no means airtight: many intermediate states exist, such as daydreaming, lucid dreaming, trances, and hallucinations (Hobson, 1999). In most dreaming states, as we have seen, the self of the dreamer is split, in a very fixed manner, into the dreamt self and observing self. This arrangement becomes looser in daydream and fantasy, but the relational you-and-me is still not in play, and perception of time and place is still weakened.

Shifting from one consciousness to another—from dream consciousness to consciousness while fully awake—presents a particularly delicate problem when using dreams therapeutically. In order to communicate about the dream, the patient must move between several states of consciousness, including a fully relational state. In the end, the dream experience has to be placed in an entirely new environment, that of the therapeutic setting, and this is by no means a given. In some instances, the dream will feel like an alien intrusion to the patient. In telling it to the therapist, the patient is unable to connect the dream material to a narrative about himself or herself.

Indications of this difficulty can be seen when the patient puts forward several versions of his or her dreamt self; or when the patient leaves his or her dreamt self out of the dream material or replaces it in favour of more anonymous protagonists. When this happens, the dream narrative tends to fragment and retrieval becomes more difficult. And even if the dream or part of it is captured, the storyline also becomes less clear and the dream material is experienced as having no apparent meaning.

In working with dreams, the therapist first of all must promote integration and access to the patient's sense of self-cohesion (Kohut, 1984; Siegel, 2010a).[17] When the dream narrative, or significant aspects of it, can be incorporated into the patient's self-narrative, elements of the dream may then serve to cue significant autobiographical memory. Therapists facilitate this process through already established narratives— that is, what they remember about their patients' essential personal information and reasons for seeking treatment (see Chapter Four).

Some patients maintain an ability to remember their dreams even after their treatment is over. They have found ways of priming their dreaming by what Hobson describes as a form of *lucid dreaming*:[18]

> By putting the notebook at my bedside and telling myself to be on the lookout for dream bizarreness, I am actively priming my mnemonic neural networks with instructions whose activation may be strong enough to persist or to reactivate when REM sleep supervenes. Enough of each activation in a localized part of my brain might then be able to resist co-option by the REM sleep activation process that would normally take over or eliminate its influence.

(2001, pp. 96–97)

Such straddling of dream consciousness and an intermediate state of awareness happens with certain patients in treatment for whom dreaming and attending to their dreams become a vital part of their therapeutic process. Particularly in a traditional Jungian analysis, these patients often find themselves remembering and recording their dreams long after their treatment ends. Their priming of REM sleep continues to provide indirect access to their self-narrative and autobiographical knowledge. With the power of such priming as a way to overcome an otherwise clear barrier between the two states, dreaming then may provide the patient with durable access to autonoetic awareness.

The uniqueness of dream stories

Dreams are unique narratives that give us important information about the narrative nature of accessing memory. Below is a summary of my conclusions about dreams as narratives and their use in treatments:

1. Dreams occur in unusual states, different from daytime consciousness. They reflect a *dreaming self*, comparable with a *wakeful self* and its capacity to create stories out of a multitude of experiences.
2. Our consciousness in dreams is primarily based on visual language. In making our dreaming experience available to our wakeful consciousness, this visual language has to be translated using verbal vehicles, many of which are metaphorical.
3. In itself, dreaming may be viewed as a dissociation. However, the tension between the *dreamt self* as the object of the dreamer's attention

and the passive *observing self* is only apparent to the dreamer after returning to daytime consciousness.

4. In most dreams, the observing self identifies with the dreamt self, such that we feel that we are dreaming about ourselves. This, in turn, makes the dreamt self function as self in *self-narratives*, or as the *working self* in autobiographical memory.

5. Patients' dream reports are primarily based on how their observing self is accessed at a later point, in a wakeful state. In conveying a dream to the therapist, the patient thus depends on an ability to remember and to build verbal bridges to the dream images.

We have seen that the ability to form self-narratives emerges gradually in a child's development. The same is true about dreaming and about the autonoetic awareness necessary for episodic memories. To narrate one's life requires three functions: (1) enough development of the left side of the brain to enable the interpreter to spin storylines; (2) a working self, to maintain autobiographical knowledge; and (3) autonoetic awareness, to reflect on past memories. Therapeutic treatments add yet another element to this mix: the capacity to translate dream imagery into verbal vehicles and to do so within a relationship especially designed for this capacity to be nurtured. Working with dreams, perhaps more than with any other image-based material, offers an opportunity to develop this capacity. In this sense, dreams are an important part of what makes treatments "narrative".

Hopefully, we will soon have more brain imaging data showing how the experience of a dream differs from the later accounting of that dream and what areas of the brain are activated in the two states. Of particular significance in this regard is locating the areas activated when one tells a dream and how they differ from those areas implicated in the dreaming experience itself. As Raymond Mar of the University of Toronto shows, there is still a considerable gap between cognitive models and neuroscientific data about narratives in general (2004). This holds true about dreams as well.

There is no doubt, in any case, that dreams can be an important part of treatment. The next chapter outlines new approaches to working with dreams therapeutically, without using the established methods of free association and amplification.

CHAPTER SIX

Metaphors and meaning

When approaching dreams as stories in the previous chapter, we discovered that they are stories that were converted from their original image language into a wide range of verbal vehicles. In the process, dreams change from being a series of moving pictures into what we may call "eyewitness reports"—accounts of a recent event that was only experienced in private. When listening to one of these reports, we realise that the concepts and metaphors must have gone through a similar transformation. Although the primary sources are clearly the dreamer's past experiences in the form of memories, expectations, fears, and desires—dreams themselves are merely wordless.

In this respect, dreams reported in psychotherapy may feel as if they were hiding their riches. Compared to other verbal media, they appear wild and incoherent, even though the images were created without hesitation while the patient was dreaming. Dreams' bizarre way of narrating may have some clinicians conclude that all we are doing, when trying to understand them, is making sense of something that in reality never did make any sense. However, if the images are examined more closely, we find that dreams use the same means of communication as those used about everyday experiences. What we are listening to is what Stephen LaBerge, a dream researcher at Stanford University, calls "a simulation of the world in a manner directly parallel to the process of waking perception minus sensory input" (1998, p. 495).

This also means that the language in dreams can be approached using our current knowledge of metaphors. Coming from a newly emerging field of research on neural patterning and preverbal thought, these findings indicate that metaphors are intimately tied to how our brain processes information. Many of the most common cognitive processes are, in fact, the result of certain neuronal groups' having been "given prior activation of other neuronal groups" (Lakoff, 2008, p. 18). This phenomenon, called *mapping*, is explained by George Lakoff, of the University of California at Berkeley and a prominent researcher in cognitive linguistics, in the following way:

> Suppose you imagine, remember, or dream of performing certain movements. Many of the same neurons are firing when you actually perform that movement. And suppose you imagine, remember, or dream of seeing or hearing something. Many of the same neurons are firing when you actually see or hear that thing.

> (2008, pp. 18–19)

From a neurocognitive perspective, our concepts of objects and movements are therefore "embodied"—shaped by the nature of our bodies and our ability to experience feelings and emotions (Johnson, 2007). And there is considerable evidence that perception of language follows similar rules. For instance, a phrase such as "he kicked the ball" appears to activate the foot areas of the primary motor cortex (Lakoff, 2008).

In psychotherapy, the metaphoric nature of what the dream narrative communicates will typically surface in the context of the session in which it is reported. But metaphors are also the means by which most communication between therapist and patient takes place. For instance, a patient may use the phrase "by sheer luck" to devalue a positive development and to guard against disappointment for fear it will not last, since LUCK IS A SUDDEN STRIKE, as in "it was a *stroke* of luck", "he has the luck of the *Devil*", and many other such metaphoric expressions and proverbs (Gibbs, 2005).[1] The therapist may then use this expression in discussing future situations and the phrase will soon take on a particular, shared relevance that differs from its initial use. The phrase has become a mutual reference to a metaphor for overcoming the fear that progress may be short-lived or for gain "success or something desirable by chance" (*American Heritage Dictionary*, 1985).

The metaphors in dreams follow similar patterns, although on a more concrete and visual level. In a patient's report of a dream, long-forgotten memories, which otherwise would have remained unconscious, may also be cued. And, conversely, as discussed in Chapter Five, certain metaphors used in the dialogue between therapist and patient may trigger the recollection of recent dreams, even without the probing of the therapist.

The established methods

The established methods for working with dreams, Freud's *free association* and Jung's *amplification*, were developed before the findings about metaphors emerged. In today's perspective, they have some obvious shortcomings. Dreams do not narrate in the same way that we tell about events experienced in the recent past; they create considerable detours when compared to the narration of other experiences. By the dream's functioning outside of regular conscious domains, the distinction between recent and remote past is not in operation and our regular attendance to perceptions of the outer world is shut down. On top of this, while having the dream, the dreamer finds his or her experiences something entirely private and is unable to relate them to others (Foulkes, 1999; Hobson, 1999).

When a patient reports on a dream about a horse, for instance, and is asked to free associate, his or her attention will focus on a wide array of material without touching upon the metaphors involved. If, on the other hand, the patient is told about instances in which this animal has a prominent place in mythologies or folktales, the neural mapping involved in the dream is likely to be lost. However, horses serve as the source for a particular metaphor, ANIMALS ARE FORCES,[2] as in the following dream reported by one of my patients:

> I had a pet horse, white, full size. We sat down on a couch or a bench. The horse sat next to me, on his rear end, his feet just touching the ground. He said to me: "There is something I want to talk to you about", but he never told me what it was. Then we got up. Both of us went into an office, felt like the waiting room to a dentist's office. People were frightened by the horse. I told them he

was friendly and wouldn't harm anyone. They let him stay in the
waiting room.

Without going into the associations to the patient's current life situation, the following metaphors are helpful in structuring a discussion of
the dream (listed with the more abstract *target* first and the more concrete or embodied *source* last). The use of SMALL CAPS indicates that the
particular wording does not occur in language as such:[3]

- ANIMALS ARE FORCES: "I had a pet horse, full size", in which the animal *horse* is mapped to the source *force* and where *full size* equals full
  force; *pet horse* then stands for the body.[4]
- PROFESSIONS ARE OFFICES: "Both of us went into an office", in which
  professions are mapped to the source *office* and the professions practised there.
- SEEKING HELP IS A WAITING ROOM: "It felt like the waiting room to
  a dentist's office"—thus, wishing to seek help is mapped to *waiting
  room*.
- FEARS ARE OBSTACLES: "People were frightened by the horse", in
  which fears of what others will say is mapped to *obstacles*.

The dream, reported several months into the treatment by a married,
sixty-six-year-old physician, indicated that there still was a central issue
that he had not been able to bring up. In the dream, this fact is expressed
in the form of a talking horse and also by going to a professional office
with the horse. This office, we must assume, is a reference to the therapy, a fact reinforced by the waiting-room metaphor.

The animal is of full size, behaves much like a human being, and
wishes to tell the dreamer something of importance. Whatever the
horse wishes to tell the dreamer may, in fact, be what the patient wishes
to tell the therapist; the dream certainly seems to reveal a wish to do so.
But the horse is also a force to be reckoned with, as the first metaphor
suggests, and the patient's fear of revealing what it involves, as the last
metaphor indicates, is considerable.

Jung's amplifications

Comparing this approach to Jung's approach of amplification—in this
case, amplifying horses to themes in mythologies or folktales—the

resulting interpretation may be similar, but the process of arriving there is quite different. When Jung interprets the dream of a seventeen-year-old girl who later died from progressive muscular atrophy, a fatal organic disease, he gives us a great deal of information, most of which may not have been communicated to the dreamer:

> "Horse" is an archetype that is widely current in mythology and folklore. As an animal it represents the non-human psyche, the sub-human, animal side, the unconscious. That is why horses in folklore sometimes see visions, hear voices, and speak. As a beast of burden it is closely related to the mother-archetypes (witness the Valkyries that bear the dead hero to Valhalla, the Trojan horse, etc.). As an animal lower than man it represents the lower part of the body and the animal impulses that rise from there.
>
> (1934a, p. 159)

In the dream, told in an initial session, the patient hears a terrible noise in the house at night. She discovers that a frightened horse is tearing through the rooms. It jumps through the hall window from the fourth floor into the street below and she sees it "lying there, all mangled" (p. 158).

Based on amplifications to myth and folktales, Jung arrives at the following interpretation: "The unconscious life is destroying itself" (p. 159). It is unclear if this is what he arrived at with the patient when the dream was reported, but his amplifications can be translated to the following four cognitive metaphors:

- THE MIND IS A HOUSE: "*A terrible noise broke out in the house at night*" thus translates into thoughts and feelings that move through the dreamer's mind at night as *a terrible noise* and these thoughts and feelings are mapped to the source *house*.
- THE BODY IS A HORSE; FEELINGS ARE ROOMS: "*A frightened horse is tearing through the rooms*" means something frightening is moving through the dreamer's body, *a horse*. Her feelings are mapped to the source *rooms*.
- JUMPING DOWN IS LEAVING ONE STATE FOR ANOTHER: "*It jumps through the hall window from the fourth floor into the street below*" means the dreamer wishes to escape her pain. *Jumping down into the street* is thus escaping what goes on in her mind, still sourced as a *horse*.

- DEATH OF THE BODY IS A MANGLED HORSE: "*I was terrified when I saw it lying there all mangled*"—the dreamer senses that her body, *horse*, will be destroyed, and become *all mangled*.

Details about the patient's personal history or her associations to the dream are missing in Jung's account. We are left with the impression that he arrived at his interpretation exclusively via his amplifications. As a result, we do not know how the patient was able to process the emotions associated with the dream. Since Jung was consulted by the physician treating the patient for her illness, he appears to have been mostly interested how to arrive at a diagnosis. To the patient, it must have felt that Jung held all the answers to her dream.

Freud's free association

Free association works in a different way, as when Freud discusses one of his own dreams about riding a grey horse and encountering a colleague, P., also on a horse. In his dream, Freud soon finds himself riding through the narrow space between two heavy wagons before dismounting in front of a small chapel with an open door. It turns out his hotel is in the same street and he is ashamed to go there on horseback. He is also shown a note of his (underlined twice), by a hotel boy. It says: "No food", and less clearly: "No work". The dream ends with "a vague idea that I was in a strange town in which I was doing no work" (1900a, p. 230).

Freud's interpretation seems, at best, partial, and he appears reluctant to spell out what he regards as the latent content of his own dream. However, his associations, which he lists in italics, indicate that the horse signifies a patient he lost to another physician.

> The horse acquired the symbolic meaning of a woman patient. (It was *highly intelligent* in the dream.) "I felt at home up there" referred to the position I had occupied in this patient's house before I was replaced by P. Not long before, one of my few patrons among leading physicians in this city had remarked to me in connection with this same house: "You struck me as being firmly in the saddle there."

> (1900a, p. 231; italics in original)

Using Freud's own associations, the horse dream has the following metaphors:

- THE OBJECT OF DESIRE IS A HORSE: "*I felt at home up there.*" Thus, the thought of the woman patient, *the object of desire*, is mapped to the source *horse*.
- MUTUAL ATTRACTION IS RIDING IN A FIRM SADDLE: Here, the confidence in a mutual attraction is mapped to *firmly in the saddle there*.
- COLLEAGUES ARE RIVAL FORCES: "*I met with one of my colleagues, P., who was sitting high on a horse.*" A colleague, P., now is thought to have a similar relationship to the patient, thus P.'s horse is also tied to the source of *forces*.
- HEAVY THOUGHTS ARE VANS: "*In this way I rode straight in between two vans.*" The source *vans*, when mapped to *heavy thoughts*, shows the dreamer, Freud, managing a conflicted state of mind.
- SHAME IS FALLING: "*It was as though I should have felt ashamed to arrive at [the hotel] on horseback.*" Still, feelings of shame overwhelm him and appear related to his body or physical feelings, of being *on horseback*.
- HAVING ENOUGH WORK IS EATING: "*A vague idea that I was in a strange town in which I was doing no work.*" Loss of work and financial support (*eating*) are associated with entering a new situation, *a strange town*.

In a later essay, Freud is more specific about the connection between symbol and symptom, between manifest and latent content. This time he discusses *hats* as symbols for the penis. He writes:

> In this connection I call to mind a symptom of obsessional neurotics by means of which they manage to ensure themselves continual torment. When they are in the street they are constantly watching to see whether some acquaintance will salute them first, by taking off his hat, or whether he seems to wait for their salute; and they give up a number of the acquaintances who they imagine no longer salute them or do not return their salute properly.
>
> (1916c, p. 340)

Here Freud concludes that the touchiness of neurotics about salutes is an unwillingness "to show themselves as of less importance than the other person thinks himself to be" and "the source of this excess of feeling can easily be found in the castration complex" (p. 340). Although he does not use the same formula for the latent content of his own dream, we have to assume that we are dealing with a similar theme. Freud must have experienced the countertransference to his previous patient as shameful and oedipal (*"I felt quite at home up there"*), which in turn could lead to the loss of new patients and being "in a strange town in which I was doing no work" (1900a, p. 230).

Shortcomings of previous methods

Free association, as the term indicates, may in the end make the patient associate to well-rehearsed stories rather than to the dream narrative itself and its images. Another shortcoming is that free association does not integrate the dream with the developing narrative of self without additional efforts by both therapist and patient. And in terms of the dream's images, as the psychoanalyst Donald Spence points out, when putting images into words, "the tendency is for the description to replace the image" (1982, p. 70). Thus, free association, by its very freedom, tends to partition the dream and leave only a distorted memory of the original experience.[5]

Amplification, in placing new and presumably comparable material side by side with the dream, appears to keep the focus on the dream itself. However, it also introduces a wide range of cultural material that may only be of tangential relevance, as when Jung mentions the Valkyries and bearing the dead hero to Valhalla. Myths, folktales, and other established narratives allow certain comparisons, but, by introducing them, what both therapist and patient attend to changes and the emotional impact of the dream is easily lost (Jung, 1961).[6] Amplification may, therefore, create an atmosphere in which the focus on individual dream images and their appearance elsewhere takes priority over exploring and integrating the dream into the patient's autobiographical narrative.

Thus, while free association encourages a semi-dreaming state by allowing whatever comes into the patient's mind to surface, amplification tends to be didactic and result in a measured and deliberate sharing of information indirectly related to the dream. In both instances, attention to the patient's self-narrative easily becomes secondary. The two existing methods also tend to promote the therapist's role as the expert and may inadvertently create a sense of his or her control of the therapeutic process.

As I discuss in Chapter Eight, some solutions to these problems are apparent in recent developments in the field; yet the two schools have basically maintained their methods for working with dreams (Bulkeley, 2001; Lippmann, 2000). Working with dreams as neurally based metaphors, on the other hand, has the potential to involve the two participants in the treatment on a more equal footing. It allows the patient's associations to the dream to produce a sense of meaning and relevance; the mutual search for metaphors keeps the discussion close to those personal associations.

This is particularly important since the telling of a dream sometimes competes with the telling of other story-based material, as when a patient reports a recently experienced event. Since most patients initially prefer to tell stories about themselves that are well rehearsed, discussing dreams may require a rather abrupt change of attitude, unless, of course, the dream can be handed over to the therapist as a random event, a story picked up in passing, without much personal implication.

This problem is less likely to occur when one is trying to establish what metaphors a dream may be built around. Dream metaphors also contain important memory cues that may not surface when one relies on associations and amplifications. As the examples in Chapters Two and Three illustrate, such cues are often the direct access to critical episodic memories; thus, to the patient's sense of self and his or her significant history.

Finally, in the two main traditions of working with dreams, the Freudian and the Jungian, often no distinction is made between the way a patient must have experienced a dream and how it is reported in the therapeutic setting. In both traditions, the focus is on the verbal account of the dream, often without attempting to reconstruct the actual experience while dreaming. As I show below, the findings about metaphors make it clear that any interpretation must first straddle the sensorimotor and primarily visual levels of the dream in order to reach the everyday level of abstraction in common language (Lakoff, 2001).

The neural mapping of metaphors

Metaphors originate in neural mapping between cognitive domains, as one of the basic activities of the brain is being connected to another, more abstract domain (Damasio, 2010). Mapping, as studied using functional magnetic resonance imaging, or fMRI, can also be seen as "the innate capacity for synapses to record and store information" (LeDoux, 2002, p. 9). To researchers using computational simulations, these activities represent an ongoing and continuous neural change, often referred to as *the generation of neural networks* (McClelland, 2000).[7] Large portions of these processes, now well documented by research, go on nonconsciously, automatically, without our awareness, and without having to be initiated or monitored. Antonio Damasio describes this basic activity in the following way:

> When certain neurons are "on", in a certain spatial distribution, a line is "drawn", straight or curved, thick or thin, a pattern distinct from the background created by the neurons that are "off". Another big difference: the lead map-making horizontal layer is stacked between other layers above and below; each main element of the layer is also part of a vertical array of elements, namely, a column. Each column contains hundreds of neurons. Columns provide inputs to the cerebral cortex (the inputs come from elsewhere in the brain, from peripheral sensory probes such as the eyes, and from the body). Columns also provide outputs toward the same sources and carry out varied integrations and modulations of the signals being processed at each locality.
>
> (2010, p. 66)

Mapping, as this description makes clear, is not static and does not result in what we think of as "maps" in conventional terms. There is no finished product, only constant motion. From moment to moment, new patterns are formed; and when some of these patterns become conscious, they are first and foremost experienced as images. Damasio again:

> The mapped patterns constitute what we, conscious creatures, have come to know as sights, sounds, touches, smells, tastes, pains, pleasures, and the like—in brief, images. The images in our minds are the brain's momentary maps of everything and anything, inside our body and around it, concrete as well as abstract, actual or previously recorded in memory.
>
> (2010, p. 70)

Although much of this imagery remains nonverbal and nonconscious and will seldom reach full consciousness, certain parts of it will become the source for what we experience as "the mind". As recent research in cognitive linguistics shows, for the latter kind of imagery to reach consciousness, some type of metaphoric or narrative mapping is necessary. Even when someone wishes to tell about an experience, such as a dream, the resulting narrative will have to include many metaphoric expressions.

Similarly, George Lakoff describes human vision as an example of such mapping:

> We are born with neural circuitry that effectively activates a "map" of one part of the brain in another part of the brain. For example, the 100 million neurons coming out of the retina grow connections before birth from the retina to other areas, including the primary visual cortex at the back of the brain. These connections form a "topographic map" of the retina in V1. That is, the connections preserve topography (relative to nearness), though not absolute orientation or absolute distance. When neurons next to each other coming from the retina fire, the corresponding neurons fire in V1 and are next to each other in V1.
>
> (2008, p. 20)

In fact, there are strong indications that metaphors are major components for the conceptual structure of language (Ramachandran, 2011). From a neurocognitive point of view, the naming of an object can be assumed to co-occur with the visual experience of this object, and firing patterns in the brain will then correlate with the neural representations of the word form. Thus, the visual experiences associated with a particular object should result in the establishment of permanent connections between these neural substrates, the actual object, and the word for it (Pulvermüller, 2003). We may also expect the same neural connectivity between motor activity and words for actions—between activity in the motor cortex and neural representations of action words (Coulson, 2008).

A longstanding belief that language functions occur exclusively in areas of the inferior frontal lobe and the superior frontal lobe, primarily in Broca's and Wernicke's areas, may therefore be out of date. Instead, the current view is that semantic comprehension involves areas of the brain previously assumed to be exclusively assigned to the sensorimotoric functions of sensing and perceiving. This would also explain why so much of our symbolic and abstract thinking is organised in metaphors that reflect visceral, sensory, and motoric activities. Especially when it comes to metaphors, therefore, we must assume that their meaning is processed in many areas of the brain and that more than one neural network is involved (Coulson, 2008).

In evolutionary terms, V. S. Ramachandran (2011) of the University of California at San Diego explains this neural function by pointing out that, as the frontal cortex expanded, areas of the brain that initially evolved as a motor cortex in mammals took on more complex behavioural functions in humans. Thus Broca's area, which controls expressive speech, sits next to the part of the motor cortex that controls the lips and tongue. Multiple pathways from the cortex to the limbic system and brainstem provide additional information of a visceral, emotional, or behavioural nature (Coulson, 2008). The left angular gyrus, for instance, appears to be involved in functions quite unique to humans, such as arithmetic, abstractions, metaphors, and word definitions. The left supramarginal gyrus, on the other hand, appears to be involved in motoric activities such as hammering a nail or waving goodbye (Ramachandran, 2011).

The language of dreams

As we saw in Chapter Five, researchers have considered many possible reasons why we dream, but without any conclusive results. On the one hand, dreaming, especially during REM (rapid eye movements) sleep, seems to be initiated in areas of the brainstem and independently of higher cognitive functions (Hobson, Pace-Schott, & Stickgold, 2003). On the other hand, when certain areas of the frontal lobes and the parietal junction are damaged, the person ceases to dream (Solms, 2005).[8] Dreaming also occurs apart from REM sleep, in what is called NREM (non-REM) sleep, which is more connected to conscious thought processes (Foulkes, 1966). Thus, what we call "dreaming" may, in fact, occur in many different states of consciousness; and what these states have in common is that they produce internally generated imagery.

What is clear, then, is that dreaming is independent of external stimuli and that specific cortical areas, necessary for processing external visual information, do not appear involved in the production of normal dream imagery (Solms, 2000). Furthermore, disorders of spoken language do not, by themselves, affect dreaming, only the ability to articulate a report of it (Solms, 1997). Thus, dreaming also appears independent of those brain regions responsible for ordinary production of language and may, in actuality, have much in common with the ability to form mental images in general (Kosslyn, 1994).

How the images in dreams find articulation in words is less clear, but memory traces from certain dream experiences—and only a small portion of them—appear available for verbal processing for no more than a short period upon awakening from sleep. As we saw in the previous chapter, most of the research has focused on establishing the overall function served by dreams and appears trapped in controversy over physiology versus psychology (Domhoff, 2005). Thus, we have limited data about how neural activity when one articulates a dream differs from neural activity when one is dreaming. Without a firsthand account of how the dream was experienced by the dreamer, the accuracy of dream interpretation is impossible to establish. We are not even sure how to determine the accuracy of the dream report, which may be one reason researchers have stayed away from questions about how dreams are remembered.

In one of the few explorations of this problem, George Lakoff outlines several possibilities for tackling a dream from a metaphorical point of view. To begin with, he feels that he can demonstrate that an interpretation, when relying on knowledge of the dreamer's history and everyday life, is based on "our everyday system of conventional metaphor" (2001, p. 169). Therefore, what may be arrived at intuitively about someone's dream most likely is grounded in the metaphoric expressions in what the dreamer reports.

The everyday system of conventional metaphors, according to Lakoff, is similar to the type of unconscious symbolism Freud called *unconscious ideation* (2001, p. 272). Whether the interpretation is made by the dreamer or another person, such as a therapist, this must be the case. Thus, when a friend who just moved in with his girlfriend tells us that he dreamt of driving through a horrendous storm and seeing the bridge to the city where he lives being blown to pieces, we will surmise that his new relationship is in trouble. This fits with what we know about his current life situation, and certain common metaphors in his dream enforce such an interpretation; Lakoff (2001) identified three metaphors specifically: EMOTIONAL DISCORD IS A STORM, LOVE

IS A JOURNEY, and RELATIONSHIPS ARE LINKS BETWEEN PEOPLE.[9] By applying the current understanding of metaphors, he therefore establishes that a dream report carries meaningful statements about the dreamer.

Mediation of meaning

According to Lakoff, there is also reason to assume that metaphors play a generative role in the dreaming itself, thus "mediating between the meaning of the dream to the dreamer and what is seen, heard or otherwise experienced dynamically in the act of dreaming" (2001, p. 272).

> Given a meaning to be expressed, the metaphor system provides a means of expressing it concretely—in ways that can be seen and heard. That is, the metaphor system, which is the place of waking thought and expression, is also available during sleep, and provides a natural mechanism for relating concrete images to abstract meaning.
>
> (p. 273)

For this second claim to be true, we have to be willing to abandon many of the psychodynamic assumptions about the nature of unconscious thoughts. From a linguistic point of view, the unconscious has a different and broader meaning as compared to Freud's and the traditional psychoanalytic view (Ekstrom, 2004). Rules of grammar and phonology, for instance, operate below a level that we could possibly have conscious access to or control over, and thus are best described as *nonconscious* (Siegel, 2003).

One explanation for this is that dreams use metaphors on what Lakoff (2001) calls the *subordinate level*. A general or conventional metaphor, on the other hand, uses as their source what he calls *superordinate* categories. A metaphor for love, for instance, may, in a dream, use the image of a particular journey as the vehicle when expressing the general metaphor LOVE IS A JOURNEY. When one is dreaming, the journey is expressed as driving a car, waiting for a train, and so on. Images of a particular kind of journey may appear, as well as, perhaps, references to a road, bridge, bad weather, and so on. The basic, fixed metaphor is, therefore, what makes the specific dream images possible. Lakoff concludes:

The fixed metaphors are fixed correspondences across conceptual domains at the superordinate level. Those fixed correspondences make it possible for basic-level imagery to have systematic meaning. Since the possibilities for basic- and subordinate-level imagery are open ended, the fixed metaphorical correspondences allow for an open-ended range of possibilities in a particular dream. Dream construction is a dynamic process that makes use of the fixed metaphorical correspondences to construct the image sequences that occur in the dream.

(p. 174)

The rules for linguistic categories are, for Lakoff, what give dream images of cars and trains a broader meaning: they refer to a general and fixed source, such as "journey". However, without incorporating deep and extensive knowledge about the dreamer's life, an interpretation is less likely to identify the specific metaphors involved in a dream. We may be able to identify general themes, such as love, work, death, and family, but those present a very large range of possibilities.

The idea that metaphors operate on different levels in dreams than in conventional stories appears to be confirmed by certain findings about the loss of the ability to dream. In his studies of 361 patients with various brain lesions, Mark Solms (1997) identified two brain locations in particular as responsible for *global anoneira*, or total loss of dreaming. The first of these, located just above the eyes and in the deep white matter of the brain's frontal lobes, contains large-fibre pathways transmitting dopamine from the middle of the brain to its higher parts (ibid.). When this site is damaged, not only does all dreaming cease, most motivational behaviour also seems to suffer (Solms, 2005).

Although damage to this pathway renders dreaming impossible, the REM cycle (the typical type of sleep for dreaming) appears unaffected. This means that we no longer can equate dreaming with REM sleep (Jus, 1973). Dreaming, as the production of internally generated imagery, in fact appears to depend on the infusion of dopamine and the activation of higher brain function in the cortex, as shown by several experiments (Hartmann, Russ, Oldfield, Falke, & Skoff, 1980; Sacks, 1990).

The second location is a portion of the grey cortex at the back of the brain, just behind and above the ears, called the occipito-temporo-parietal junction (Solms, 2000). Some of the highest levels of perceptual processing occur in this site by converting concrete perceptions into abstract thought and memories.[10] Since this process may be reversed when it comes to internally generated imagery, dreams can be understood as "projecting information backward in the system", thus "being fed backwards into the cortex as if they were coming from the outside" (Kosslyn, 1994, p. 75; Zeki, 1993). Internally generated images, such as vision, mental imagery, and so on, appear to be processed differently than sense perception but, nevertheless, via similar neural pathways.

These findings appear to dovetail with what Lakoff suggests about metaphors in dreams: they operate on a subordinate level that is concrete and specific, even when the dreamer later describes them verbally, on a superordinate level. One of the therapist's roles in treatments, when knowing the patient's personal history and the associations to the patient's dream images, may, in fact, be to assist in the patient's process of articulating the more abstract, everyday level of the dream metaphors presented. This process does not require an exclusive focus on far-reaching associations, nor does it demand extensive amplifications. The discovery process will be a mutual one between the therapist and the patient, while still staying with the meaning offered by the dream.

The grammar of metaphors

This discovery process assumes that dreams share much of their grammar with metaphors. That this is the case will become apparent when the patient remembers the specific form of narrative used by the dreaming self and is able, as faithfully as possible, to translate the visual and concrete metaphors as they were experienced while he or she was in dreaming consciousness.

Researchers working with lucid dreamers—persons who are able to stay conscious while dreaming—have, in fact, recorded aspects of this dreaming consciousness. According to Stephen LaBerge, the visual metaphors in their dreams are reflected in the dreamer's brain and body:

> Using these approaches (eye movement signals from lucid dreamers plus physiological measures) in a series of studies ..., my colleagues and I discovered that various dreamed experiences (e.g., time estimation, breathing, singing, counting, and sexual activity) produce effects on the dreamer's brain (and to a lesser extent, on

the body) remarkably similar to the physiological effects that are produced by actual experiences of the corresponding events when the subject is awake.

(1998, p. 501)

In this sense, the dream shares features with what happens in other instances of nonlinguistic communication. In films, a metaphor such as FALLING IN LOVE IS PHYSICAL FALLING may be expressed as cascading down a waterfall; the metaphor SEXUAL DESIRE IS FIRE is often dramatised through a fire in a fireplace burning nearby the amorous couple (Kövecses, 2010, p. 63).

Similarly, the metaphors in dreams remain visual and concrete. Rather than the abstract and conceptual level used in most conventional metaphors, dreams must use metaphors on a basic and nonverbal level (Lakoff, 2001). LOVE IS A JOURNEY, for instance, is a metaphoric expression requiring words. We would have difficulty expressing it by exclusively visual means; a more concrete form would likely be required, as in the case of "arriving in Paris, the city of love", in the dream discussed in Chapter Five (see also below).

Certain metaphors may also be acted out. Zoltán Kövecses of Eötvös Loránd University, Hungary, writes:

> A major conceptual metaphor for difficulty is DIFFICULTIES ARE BURDENS. Sometimes people do "act out" this metaphor, when they walk in such a way that suggests carrying a heavy load on one's shoulders. In these cases, physical symptoms can be seen as "enactments" of conceptual metaphors. A large part of learning the profession of acting involves learning how to act out certain conceptual metaphors.
>
> (2010, p. 64)

In dreams, these enactments are often felt as bodily events happening to the dreamt self, not as playful and voluntary acts. In this sense, they are more like the type of enactments that have received attention in current therapeutic writings, when a patient unconsciously reenacts often-traumatic memories in the treatment (Ginot, 2007; Schore, 2003).

The level of expression for a given metaphor thus depends on what brain sites are available for processing. And since this choice occurs nonconsciously, the memory traces activated lack chronology and separation, even though the common mechanism for generating metaphors remains. This also means that dreams, to a large extent, are created in an environment devoid of our everyday semantic memory of word meanings, categories, facts, and propositions (LaBerge, 1998). Thus, we find it much harder to remember what was told to us in a dream than what we tell someone about experiences in a wakeful state. We have to rely on our episodic memories reflected in the dream; but also, here, we are faced with the difficulty that the normal chronological markers are missing. The dream will introduce both recent and very old memories, sometimes in the same sequence, sometimes even in the same character.

Since metaphors work on an unconscious level, therapists wishing to integrate how metaphors appear in dreams may not always be able to pinpoint how a particular metaphor cued the patient's associations and memories represented in the dream. By focusing on the dream's metaphors, however, the patient's autonoetic awareness may be activated and what is retrieved would add to the patient's developing self-narrative.

This way of translating the dream is, naturally, only one of many possible approaches. Interpreting a dream remains similar to translating poetry from one language to another, or from poetic expression to prose. However, the research on the neural basis for metaphor makes it possible to look for this underpinning without having to fall back on the established methods of amplifications to other narratives or on a rigid formula for what the dream wishes to accomplish—the legacies of Jung and Freud, respectively.

The visit-to-Paris dream

When we examine the dream (cited in the previous chapter) of a thirty-two-year-old patient entering analysis, several metaphors become apparent. Some of them became evident after I heard the associations my patient had to his dream; others emerged in connection to information he later gave about his current life situation.

Paris, which he had never visited, is, in his imagination, a place of passion and love. He is aware that the hill in the centre of the city does not exist in reality, and initially he was surprised by what was going on there. Most confusing to him were the many bearded older men and the funeral, but he agreed that this could reference the fact that he grew up without knowing his father (his parents divorced when he was one year old and his father moved out of state). By the time he was in his twenties, he had met his father for the first time. However, through his becoming interested in genealogy, he had studied his father's family and already made contacts with several members.

It also became clear in the sessions following the dream that the impetus for his seeking treatment was a relationship that had begun nine months earlier. The patient was now living with his girlfriend, but they were having frequent fights and he was beginning to doubt that the relationship would work out. He had difficulty finding common ground with his girlfriend and living together was a struggle; thus, the hill in the dream may be a metaphor expressing his current situation of "an uphill battle" to maintain his original sense of hope of a lasting intimacy.

At the time of his telling the dream, many of the episodic memories attached to the images were still not fully explored. A significant element is the fact that a major portion of the dream is told from a narrator's point of view, and only the last scene involves a dreamt self. In fact, the dream may be viewed as having a dream-within-the-dream, with the dreamer himself as the main character only at the beginning and the end. In spite of this, the narrative is quite clear; it may, in many respects, relate to love and poetry, but also to journey, effort, and renewal.

By including the associations the patient provided, we arrive at the following version of the dream:

> To my surprise I am visiting Paris, the city of love [living with girlfriend but having frequent fights and difficulty finding common ground with her]. I am discovering an elevated new part of this city where old men in long beards are meeting for what appears to be a funeral [grew up without knowing father]. They are also celebrating the return of a famous folksinger, a poet and their hero [writing is very important to me, but I have been blocked and am uncertain if this is what I want to do]. He has been held captive by Islamic terrorists [I grew up with mother and grandparents who often had violent fights]. He has learned Arabic and can communicate with people in the Arab world. He brings several children born in captivity. I am climbing this hill and must hold on to the side rails.

This leaves us with at least ten metaphoric expressions, many of them primary metaphors with time-related domains as the target:

1. LOVE IS A JOURNEY OF SURPRISES. *"To my surprise I am visiting Paris, the city of love"* is how the dream presents the patient's feelings about the current relationship as being part of his *life journey* (Lakoff & Johnson, 1999; Lakoff & Tuner, 1989).

2. KNOWLEDGE IS SEEING; MORE IS UP. *"I am discovering an elevated new part of this city"* shows that he is discovering something new in this relationship that is more than in previous situations (Lakoff, 2001).

3. FATHERS ARE BEARDED FACES. *"I see a gathering of old men in beards"* points to a need for father figures that were missing as he grew up (Kövecses, 2010).

4. DEATH IS DEPARTURE; FUNERALS ARE GOODBYES. *"A meeting for what appears to be a funeral."* Both metaphors seem to point to a need to let something die.

5. NEW LIFE IS RETURNING HERE. *"Return of a famous folksinger, a poet."* Another positive metaphor about the patient's interest in being a writer.

6. TERROR IS PHYSICAL FORCE; CAPTIVITY IS CONTAINMENT. *"The hero has been a captive of terrorists."* Both metaphors point to a traumatic childhood (Johnson, 2005).

7. COMMUNICATION IS FOOD; STATES ARE LOCATIONS. *"He can communicate with people in the Arab world."* Another indication, in the forms of metaphor, to communication and being able to find expression for his problems.

8. FUTURE IS AHEAD; NEW IS CHILD. *"He brings several children born in captivity."* The sources of these metaphors are *ahead* and *child*, thus the sense of potential developments.

9. EFFORTS ARE HILLS. *"I must climb this hill."* The dream moves into first person and appears to indicate that *efforts*, the target of *hills*, are required.

10. HELP IS SUPPORT. *"I must hold on to the side rails."* The side rail relates to the source *support*, perhaps what he is looking for in the therapy (Lakoff & Turner, 1989).

The first metaphor appears to stress the dreamer's current situation and relates it specifically to his nine-month-old love relationship (metaphor 1). The next set of metaphors occurs in the part of the dream in which he is only the observer, not the *dreamt self* (metaphors 2, 3, 4, 5, 8). This part of the dream is the longest and begins with references to discovery and elevation. The metaphors KNOWLEDGE IS SEEING and MORE IS UP (metaphor 2) tell us about the many things he has discovered in his new relationship, including the painful reliving of the relationship with his mother. Next come several metaphors related to death and new life (metaphors 3, 4, 5). In FATHERS ARE BEARDED FACES, a part of the human anatomy is the whole person and maleness is identified with beards in a child's mind.

The second surprising discovery emphasises that DEATH IS DEPARTURE and FUNERALS ARE GOODBYES and the need for closure. Metaphor 5, NEW LIFE IS RETURNING HERE, may originate in metaphoric expressions related to being alive as a state and birth as a change of location (Lakoff & Turner, 1989). The combined metaphors TERROR IS PHYSICAL FORCE and CAPTIVITY IS CONTAINMENT (metaphor 6), are poignant references to past terror and abuse. The source concept of "captivity" is "containment", as in the expression "he *was held* by his captors". The source concept of "physical force" can be found in many expressions related to person-to-person violence and terror, such as "she was *forced* into submission", "the police provided *enforcement*", or "*physical force* had been inflicted on the child". Especially when related to force inflicted by another person, such metaphors may reveal a sense of injury and experiences of traumatisation (Kalsched, 1996).

The heroic struggle to articulate these experiences is then portrayed as new beginnings (metaphor 8), as in expressions such as "*my children are my future*". The scene describing how the poet has learned Arabic and is able communicate with people in the Arab world appears to combine two metaphors, COMMUNICATION IS FOOD and STATES ARE LOCATIONS. Thus, there are indications that the patient's problems with his girlfriend involve a lack of meaningful communication and also that his chauvinistic attitude about the problems they face is, in his mind, related to Arab attitudes toward women.

The last two metaphors, again, involve the dreamer directly and the dreamt self. EFFORTS ARE HILLS (metaphor 9) points to the general sense, when entering treatment, of a long and uncertain struggle; while the last metaphor, HELP IS SUPPORT, may speak to the patient's realisation that he needs help.

The identification of these metaphors served many purposes. The initial dream gave me access to several lines of communication: the metaphors cued significant ongoing and past experiences that together we could explore in the following sessions.

As to the structure of the dream, one feature stood out to me immediately. The dreaming self, as defined in the previous chapter, is here primarily the dream's narrator—not its main character. For much of the dream, my patient dreams of himself as an observer rather than as an actor or participant. Instead, the dream action is carried out by some old men in beards and, to some extent, by the French folksinger, who seems to represent a personification of the patient's ambitions to be a writer and poet. The dreamt self, which appears only at the beginning and end of the dream, therefore represents the patient's immediate sense of self, his actual and current experience of self. This split may, in fact, represent a dissociation between this current self and his experiences of terror and captivity in the past. However, in spite of the divided narrative, the dream has a clear beginning, turning point, and ending.

The dream narrative tells us that the patient finds himself in the new and confusing environment of trying to build a secure, intimate relationship with his girlfriend. Many unresolved issues from the past do surface and the dream deals with them as a dream-within-the-dream— as if this review of the past were inserted in the main dream. When the narrative again returns to the present, and with the patient himself as the dreamt self, it tells us of his having to work hard and of his need for support to accomplish this.

Types of metaphors

To cognitive linguists, the main characteristic of metaphors is the neural mapping of a concrete concept to a more abstract one. By using two categories, a sensorimotor domain (sense perception and motor activity) and a domain such as a subjective judgement, metaphors connect what the body does with the notion of achieving a purpose (Kövecses, 2010).

In my patient's dream, for instance, in the last metaphor, HELP IS SUPPORT, the source "support" is based on a common physical activity. The experience of offering (or being offered) support implies assisting (or being assisted) physically—for example, when confronted with difficulty in walking due to an injury.[11] In the dream, since the activity has to be expressed visually, the support comes from side rails, a more concrete representation of the kind of help the patient is seeking now. This is consistent with the dream's using a subordinate level of expression for the metaphor, rather than the common and abstract level in HELP IS SUPPORT.

In common metaphors, the result of mapping one domain to another is a *primary metaphor* (Lakoff & Johnson, 1999). In the metaphor AFFECTION IS WARMTH, a subjective judgement (affection) adds new dimensions to a sensorimotor domain (temperature) in a one-way inference.[12] The primary experience, the sensation of warmth, is tied to the experience of being held affectionately. Although the two words come from different cognitive domains, the metaphor is what joins them, in what Lakoff and Johnson call *source to target*—"warmth" being the source, and "affection" the target (see also Johnson, 2008).

Regarding the source of this metaphor, "warmth" relates what we perceive via our senses. Other sensory sources are "light and darkness", "cooking and food", and "health and illness". "Forces", "movement and direction", "animals", "plants", "buildings and construction", and "machines and tools" relate to body and movement and to our capacities for vision and negotiating space. Thus, the most common form of action, moving the body, is here used to characterise aspectual structure, the structure we find in actions and events in general (Lakoff & Johnson, 1999).

The most common target domains, on the other hand, are "emotion", "desire", "morality", "thought", "society/nation", "politics", "economy", "human relationships", "communication", "time", "life and death", "religion", and "events and actions"—more abstract and conceptual domains (Kövecses, 2010).

Primary metaphors may also be the foundation for what Lakoff and Johnson (1999) call *complex metaphors*. Here, forms of commonplace knowledge, such as cultural models, folk theories, or "simply knowledge or beliefs that are widely accepted in a culture" are the targets (p. 60). As an example of such a metaphor in our culture, they mention A PURPOSEFUL LIFE IS A JOURNEY, since in our culture "people are supposed to have a purpose in life, and they are supposed to act so as to achieve those purposes" (p. 61). The primary metaphors, PURPOSES ARE DESTINATIONS and ACTIONS ARE MOTIONS, thus are at the core of a more complex metaphorical version of that cultural belief. The complex mapping found in such metaphors is:

A PURPOSEFUL LIFE IS A JOURNEY
A PERSON LIVING A LIFE IS A TRAVELLER
LIFE GOALS ARE DESTINATIONS
A LIFE PLAN IS AN ITINERARY.

Metaphors and image schemas

In explaining the power of metaphors, Lakoff, in his collaboration with Mark Turner of Case Western Reserve University, points out that certain concepts, such as "journey", also are knowledge structures they call *image schemas*. Once learned, such skeletal structures can be used "automatically, effortlessly, and even unconsciously" (Lakoff & Turner, 1989, p. 62). In the metaphor LIFE IS A JOURNEY, "journey" is the source, but it has other well-differentiated components, such as "traveller", "a starting point", "a path", "impediments", and so on, that can be fitted into available slots. Those slots, such as a particular person who is on a journey—a "traveller"—are thus another version of the same metaphor. In understanding that someone is a traveller, we also understand that he fills the role of "traveller" in a journey schema (p. 62).

Such schemas have recently received considerable interest also among developmental psychologists in that they appear to be what Mark Johnson calls "embodied origins of human meaning and thought" (2007, p. 29). As deeply held patterns of experience, these schemas were likely to be instantiated in our nervous system. Tim Rohrer of the Colorado Advanced Research Institute summarises the thinking about image schemas:

> The notion of an image schema may have originated in linguistics and philosophical hypothesis about spacial language, but—given the recent evidence from cognitive neuroscience—is likely to have its neurobiological grounding in the neural maps performing somatomotor and multimodal tasks.

> (2005, p. 172)

At the beginning, the list of these schemas was fairly short and based primarily on bodily perceptual experience (Grady, 2005). Aside from the "journey" schema of *source-path-goal*, Johnson and Lakoff (1980), the originators of the concept, mention *containment/container, part-whole, centre-periphery*, and *balance*, as well as several schemas for *force*.

Johnson, in his original definitions, also stressed the dynamic nature of these schemas: they are "flexible in that they can take on any number of specific instantiations in varying contexts" (1987, p. 29).

The schema *containment/container* seems to originate in a sense of something being on the outside or the inside. When projected, this means that cups, rooms, and houses are containers, all common occurrences in dreams. We also tend to impose boundaries on what we perceive, thus marking territory so that it has an inside and a bounding surface—a wall, a fence, or an abstract line or plane to create a sense of *in* or *out* (Lakoff & Johnson, 1980). Thus, when one sees a container, or hears or reads the word "in", the *container* schema will be activated. Johnson writes: "Certain types and sizes of containers will offer different specific affordances for a being with our type of body, brain, and environment" (2005, p. 22).

Robert Dewell (2005) of Loyola University, in stressing the dynamic aspects of *containment*, focuses on it as a universally important semantic concept that can be found in much inferential reasoning and in metaphoric structuring. He also finds it preverbal in that a child has the subjective experience with things going in and out of containers and seeing one object containing another, even when not involved physically in the act.

The schema *part/whole* most commonly is found in a metonymy in which one entity is being used to refer to another, as in "I am reading Shakespeare" or "Nixon bombed Hanoi" (Kövecses, 2010). Thus, the PART FOR THE WHOLE metonymy is based on parts that are generally related to the whole. When we say that "we need some new faces around here", we mean not only faces but whole persons. And when we say "they have many good heads in their organisation", we mean that they have many people working there who are intelligent. In these references, we're using the *part/ whole* schema (Johnson, 1987).

The image schema of *force*, finally, is a central schema for metaphors describing violent bodily action, as in the dream above about terror, which was translated into the metaphor TERROR IS A PHYSICAL FORCE (Johnson, 1987). *Force* is, therefore, an image of what "acts upon us", as the dreams noted earlier in this chapter show, and they often manifest in images of animals, in what George Lakoff calls "the basic (and subordinate) level" (2001, p. 274). The schema for *force* can also be found in such references to sexuality as "sexual vibes"—thus the metaphor SEXUALITY IS A PHYSICAL FORCE, described by Lakoff (1987).

Prelinguistic representations

As basic knowledge structures, image schemas open new ways of understanding preverbal learning; as knowledge structures acquired very early in a child's life, they must be due to what researchers call *conflation*, the lack of distinction between two conceptual domains. Lakoff explains:

> Image schemas structure our experience independently of language. For example, we experience many things as *containers*—boxes, cups, baskets, our mouths, rooms, and so on. Prior to learning language, children go through stages of exploration in which they repeatedly put things in and take them out of many different kind of objects, thus treating these objects as containers.
>
> (1987, p. 60)

The fact that image schemas can be traced to early childhood cognitions also makes them prelinguistic. Jean Matter Mandler, who researches cognitive developments in infants at the University of California–San Diego, defines them as "special representations that express primitive or fundamental meanings" (2004, p. 78). She claims that image schemas are in place before the child's first birthday and are the building blocks that the infant uses to "form images, to recall, and eventually to plan" (p. 91).

Thus, from an early age, this perceptual system "parses and categorizes objects and object movements (events)" (ibid., p. 91). Mandler explains:

> We appear to think in sentences, whose components are concepts couched in propositional format. Not only is this phenomenal experience of thought equivalent to language at least somewhat illusory but also it is of no use in understanding prelinguistic representation. My introspection says I think in English sentences, but is a Korean baby going to think in English sentences? Obviously not, nor in Korean sentences either. What is needed is a more universal representational format, one that is not specifically language based but still suitable for learning whatever language one's native tongue will turn out to be.
>
> (p. 77)

According to this understanding, the infant learns to identify particular kinds of things well before they are able to use language and create stories to communicate these schemas. In turn, however, these early schemas enable a wide range of new concepts to develop (Mandler, 2004).

That early childhood cognitions are universal and may hold the clues to later developments is not a new thought. Freud, in articulating theories about childhood amnesia, assumed that "the material of the forgotten experiences of childhood is accessible to dreams ... with all its characteristics, it egoism, its incestuous choice of love-objects, and so on" (1916–1917, p. 210).

Freud's insights into what we today may consider the early development of certain image schemas were not based on research on infants but on his experience of adult patients' dreams. However, his knowledge of the memory limitations of infants allowed him to formulate the theory that dreams are a form of regression into the archaic world of infancy. In his "Introductory Lectures on Psycho-Analysis", this also means that the infant's choice of an incestuous love-object is "the first and invariable one", only later to be resisted (1916–1917, p. 210).

While childhood amnesia, to Freud, offered hypothetical answers about how attachment patterns and later sexual orientation are affected by the relationship to caregivers, the findings about image schemas have provided many answers about what is fundamental to the development of language. Turning again to Mandler, many image schemas appear to be in place "before the particulars of culture have much opportunity to bias conceptualization" (2005, p. 138). She concludes:

> Until recently we had almost no information of preverbal concepts. In part because of the influence of Piaget's view of infancy as purely a sensorimotor period, it was largely assumed that infants have no conceptual life, and so it went unstudied. This has changed dramatically in the last decade or so. We have learned that a number of high-order cognitive functions develop in the first year, as recall, making inductive inferences, and mental problem solving.
>
> (p. 139)

Similarly to Freud, Jung hypothesised that certain dream images were "the products of the primitive mind, its *representations collectives*, or mythological motifs" (Jung, 1961a, p. 228). Such archetypes, he writes, are "an inherited *tendency* of the human mind to form representations of mythological motifs—representations that vary a great deal without losing their basic pattern" (ibid.; emphasis in original).

The idea of archetypes has recently been revisited Jean Knox (at the University of Kent), a Jungian psychoanalyst. In her book *Archetypes, Attachment, Analysis* (2003), she argues that, based on research on image schemas, archetypes may be understood as representations of "the interpersonal processes which form the foundation for the construction of meaning in the mind" (p. 206). In agreement with Mandler, she views these processes as present in the human infant from about the age of six months:

> This is a two-way process; regularly repeated patterns of experience become internalized and so form part of our psychic structure as mental models and these in turn structure our perception of the world, determining how we interpret what we see, hear, and feel. The repeated patterns of daily experience gradually build up into a generalized schema which forms the basis, for example, for the symbolic concept of mother or father.
>
> (p. 206)

Research on image schemas thus appears to touch upon core elements in the construction of metaphors. Attempts to capture these elements, however, may also be found in Freud's formulations about dreams as a regression to "this infantile level" into which "dreams carry us back every night" (1916–1917, p. 210). In Jung's formulations, they seem to be conceptualised as archetypes. And in the work of John Bowlby (1969), the founder of attachment theory, image schemas appear to be thought of as *internal working models* (also called *mental models* by later theorists such as Fonagy, Steele, Moran, Steele, & Higgit (1991)).

Dreams and treatment

As I have described in previous chapters, findings in neuroscientific research, when applied to therapeutic formulations, make episodic memory and narratives critical to our understanding of what happens in treatments. And as for psychodynamic approaches in particular, new data about the function of metaphors now open fresh avenues to working with dreams.

In metaphors, two facilities are called on at once: on the one hand, perception or motor control; on the other hand, categorising and reasoning. Metaphors create links between deeper, unconscious structures and the more abstract ones that we associate with consciousness. They do so, according to Lakoff and Tuner, "automatically, effortlessly, and even unconsciously" (1989, p. 62). Thus, in identifying metaphors in dreams, the therapist's own biases are less likely to interfere than when the therapist uses free association or amplification to develop understanding of the patient's dreams.

Metaphors were enigmas to Freud and Jung. Their understanding of language, concepts, and abstractions was, to a large extent, based on the current science. At that time, the philosophical outlook treated metaphors as cognitively insignificant (Johnson, 2007). That sense perception and motor activity are the sources for neural mapping would have been impossible to articulate before recent research.

In his review of previous notions about metaphors in Western traditions, Mark Johnson (2007), the philosopher who coauthored many presentations on metaphors and image schemas with George Lakoff, points to an objectivist-literalist view as the main culprit for this enigma about embodied cognition. According to a long-prevailing view, "a fixed and determinate mind-independent reality, with arbitrary symbols that get meaning by mapping directly onto that objective reality" was beyond doubt (Johnson, 1987, pp. xxi–xxii). Reasoning, it was held, "is a rule-governed manipulation of these symbols that gives us objective knowledge, when it functions properly" (p. xxii).

In many ways, psychoanalysis represents the first major break with this objectivist-literalist tradition. When dealing with dreams, Freud and Jung came to the realisation that what their patients told them about their nighttime experiences was only a linguistic expression of underlying conceptual and unconscious structures. They called these structures *symbols* (Freud, 1900a; Jung, 1913a). The solutions the two men arrived at may have differed, but both tried to find a method for reaching the underlying and unconscious building blocks of dreams.

Today, we have additional and more precise tools to unveil these unconscious structures. Working with dreams may not be the solution in all treatments, but the narrative and metaphoric expressions that are so typical of dreams give us a unique entry into psychological areas of personal meaning. Such access will often enhance the participation of both patient and therapist in the therapeutic process.

PART II

REMEMBERING, REPORTING, AND TEACHING

Parents, educators, and therapists—those of us who should be most concerned with shaping minds—usually pay little attention to the brain. I have heard therapists say that psychotherapy is an art and that the brain is irrelevant to their work. I would respond, as with any art, a thorough knowledge of our materials and methods can only enhance our skills and capabilities. The brain is a treasure trove of information about where we have come from, what we are capable of, and why we act as we do. It holds many secrets about how we can know ourselves better and improve the way we do psychotherapy, teach, and parent our children.

—*Louis Cozolino*, The Neuroscience of Human Relationships
(2006, p. 9)*

*The epigraph to Part II is taken from *The Neuroscience of Human Relationships: Attachment and the Developing Social Brain* by Louis Cozolino. Copyright © 2006 by Louis Cozolino. Used by permission of W. W. Norton & Company, Inc.

CHAPTER SEVEN

Where it happens and how

Most of our knowledge about what happens in psychotherapy comes from what therapists describe about treatments they participated in. Aside from taped recordings of sessions used for outcome research, this knowledge is in the form of reports, progress notes, and other documentation that was created after the fact (Messer & Wolitzky, 2007).

When considering what therapists report, we are therefore dealing with past events and how these events were remembered by one of the participants in a two-way dialogue. Even in notes taken shortly after a session or when otherwise describing the therapeutic process, therapists must use some type of narrative in order to remember and organise what they experienced. Narratives are therefore the basis for how therapists report on their work and what they remember about it. In fact, much of the ongoing communication with a patient relies on how well these narratives can hold memories for what occurred in previous interactions.

In the first part of the book, I discussed how narratives are constructed and how they tell us who we are, what we remember, even how we dream at night. In this part, I more closely examine the types of memory that therapists develop in order to serve a healing function for their patients. Using current research and material from outcome studies, I describe six particular forms of memory. Two of these relate to the therapist's episodic memory of sessions with a particular patient, two are based on what the therapist brings to each treatment, and two are of a more general nature and are based on certain learned skills and approaches. (See also Table 4.2 in Chapter Four.)

I will also shed some light on the processing that occurs in the here-and-now between therapist and patient—the short-term perception, rehearsal, and encoding of what the other party communicates. When considering the different forms of data that both therapist and patient have to process, it is no surprise that what each maintains in long-term memory is quite limited.

Mutual activation of memory

We may never be able to document fully, from session to session, how healing takes place in a patient; however, we must assume that the effectiveness of the therapist's communication is a major factor. Patients absorb and eventually integrate into their lives outside the therapeutic setting how the therapist uses memory. Patients learn from how their therapist remembers; and what their therapist remembers will activate further recollection, which in turn helps their therapist understand more aspects of his or her patients.

This is, thus, a reciprocal and dyadic process. Psychotherapy treatments are at their core a mutual activation of memory. The power of the process cannot be explained in any other way, especially if we consider how much of the details from treatments are lost from session to session, much of it within seconds or minutes. To therapist and patient alike, loss of much of what transpires in a session increases rapidly from the first hours after that session (Spence, 1982, p. 231).

When therapists describe interactions that took place in specific sessions, they often do so as if they were happening in the here-and-now. But since we are dealing with reports of past events, therapists can only describe what they remember about those events. However, even in outcomes studies, we have no exact numbers for what the average therapist remembers about a session, whether that be shortly after it ended, the next day, the next week, or several months later.[1]

We must assume that some therapists are particularly skilled at remembering the immediate dialogue from sessions and are better at this than other therapists. In fact, individual differences must exist among therapists in terms of the kinds of skills they use. Some therapists probably do a better job encoding into long-term memory what transpired and are better able to consolidate their memories into episodic memory. Others probably do better remembering semantically relevant information about their patients, such as profession, age, place of birth, and the names of parents and siblings.

We also know that different treatment methods—for instance, cognitive versus psychodynamic approaches—focus the therapist's attention in different ways, thus emphasising different types of skill sets. All these factors influence how the individual therapist remembers (Teyber & McClure, 2000).

What all therapists have in common, however, is that their recollection of what happened in a session decreases rapidly a short period of time after the session ends. Therefore, we must also assume that some of the information supplied by the patient never enters the therapist's long-term memory and thus is never encoded. So when a therapist writes a case report involving sessions that took place many months earlier, major portions of what transpired have, by then, been permanently lost (Baddeley, 2000).

The role of attention

When we turn to what neuroscientists have to say about memory, the first thing we discover is how dependent it is on attention and how attention will fluctuate widely within a session. Even for the most immediate details of what we deal with and the events we participate in, there are both external and internal stimuli capturing a therapist's attention. Michael Gazzaniga of the University of California at Santa Barbara explains this phenomenon as follows:

> Only certain information makes it through to consciousness. It is a dog-eat-dog world in our brains. Experiments have shown that in order for a stimulus to reach consciousness, it needs a minimal amount of time to be present, and it needs to have a certain degree of clarity. However, this is not quite enough. The stimulus has to have an interaction with the attentional state of the observer.
>
> (2008, pp. 268–269)

When attention is consciously directed, neuroscientists call this *top-down processing* (ibid., p. 286). However, sensory signals from something sudden and unexpected may also capture the person's attention and co-opt conscious control. In such situations, the reverse occurs, a *bottom-up processing*. Attention is therefore controlled by many cortical processors beyond our conscious control. And, as we will see in discussing certain executive functions, below, the distinction between these two processing modes makes it clear that attention and consciousness are what Gazzaniga calls "two separate animals" (ibid.). What is perceived in a given moment will only later become encoded into long-term memory and most of what we perceive from moment-to-moment will be discarded when not immediately put into use.

Joseph LeDoux of New York University captures a basic fact about how short-term memory functions (referring to discoveries in the 1970s by Alan Baddeley and Graham Hitch):

> How many times have you looked up a phone number and then forgotten it after being momentarily distracted? The reason for this is that you put the number into working memory, a mental workshop that accommodates one task at a time. As soon as a new task engages working memory, the content of the old task is bumped out. For that reason, unless you keep rehearsing the phone number and manage to ignore other things that compete for your attention, it will not remain in your mind.

(2002, p. 175)

What neuroscientists call *working memory* is therefore what functions in the immediate circumstances of a psychotherapy session. A highly sophisticated system of buffers and executive functions, working memory allows therapists to keep in mind what someone is saying while also observing the person's body movements and facial expressions. Working memory, thus, has buffers for holding verbal as well as nonverbal information, but only temporarily—perhaps for seconds. Specialised buffers are present for sights, sounds, smells, and other sense perceptions. Buffers also exist that hold the subject of each sentence someone utters until a verb appears; and when someone uses a pronoun whose referent is in a previous sentence, we have the ability to consult what is held in that verbal buffer. We may even have a buffer for remembering a very recent event long enough to be able to tell about it (LeDoux, 2002).

Short-term slave systems

In the initial description by Baddeley and Hitch, working memory consisted of two domain-specific buffers that were later termed *slave systems*. Further studies have expanded this conception to include other buffers as well. Researchers have also placed more emphasis on the role long-term memory serves in the workings of executive functions (Baddeley, 1986).

For our purpose of understanding what transpires in the psychotherapy setting, the most obvious buffer has to do with verbal comprehension and the temporary storage of phonological information. This auditory buffer holds the sounds of the patient's speech automatically by entering them into the therapist's phonological loop. Within seconds, they are then rehearsed subvocally and stored long enough for the therapist to encode an entire sentence. The same is true for the patient's processing of the therapist's speech.

The more time it takes to pronounce a word, the more time it takes to rehearse it. Also, words that are phonologically similar require more attention than those that have distinctly different sounds. For instance, *cat, rat, mat* take more effort to rehearse than *man, egg, boat*, although they are of the same length (Gathercole, 1997). Should irrelevant sounds disturb one's attention, the rehearsal becomes more complicated. It also appears true that we have an easier time rehearsing genuine words than meaningless combinations of sounds. This effect is due to familiarity with the sounds of the word rather than due to its semantic meaning (Baddeley, Thornton, Chua, & McKenna, 1996). However, when the phonological structure of a word is familiar, the latter also appears to be more easily rehearsed (Gathercole, Conway, Collins, & Morris, 1993).

Neuroanatomically, the phonological loop appears to activate parietal regions of the brain, usually in posterior, opercular, and prefrontal motor regions (Nyberg & Cabeza, 2000). According to functional neuroimaging studies, the rehearsal process also activates Brocca's areas (ibid.).

The face-to-face information

Not only does remembering by putting sound bites together require complex mechanics, but, for the material to be well encoded, we must also interpret what we are hearing. And since they are listeners by training, therapists must also rely on other channels of communication to cue their attention. If we include what is expressed without words in a given session, the amount of information being transmitted in that session is greater by far than what one might otherwise assume.

According to the working-memory model, speech has a somewhat preferential status compared to written communication. Speech does not need to be converted from printed words and thus requires less mental effort to be remembered. When we read a text, our attention is more concentrated within a narrow field of perception, leaving us less room to attend to nonverbal processing (Baddeley, Thornton, Chua, & McKenna, 1996).

For several reasons, visual and motor information—often called *nonverbal communication*—take on particular importance in the therapeutic encounter (see Table 7.1). Therapists are generally trained to respond to information patients offer via their movements, gestures, pauses, and tones of voice. These channels are no doubt at work in the background of most human interactions. In the therapeutic encounter, they take on additional significance. Via the patient's nonverbal communication, therapists process affective experiences that their patient may not be able to communicate verbally, or simply are unaware of.

In Baddeley's model for working memory, nonverbal information is stored in a buffer he calls the *visuospatial sketchpad* (Baddeley, 1986). This buffer appears to be anatomically dispersed, with visual data mostly involved with bilateral occipital components and spatial data more involved in parietal regions. The two components of the buffer are, in fact, easily dissociated from each other, with object information typically found in occipital-temporal and inferior prefrontal regions. Activation related to maintenance of spatial information, on the other hand, is usually observed in occipital-parietal and superior prefrontal regions.

How therapists process information from the visuospatial buffer while also attending to explicit verbal communication is by no means clear and must differ from therapist to therapist. No studies address the working memory in therapists; moreover, the clinical literature uses different models to describe it. Roy Schafer, for instance, understands nonverbal communication as related to transference; he claims that what he describes as "bodily rigidity, lateness to or absence from scheduled sessions, and mumbling" are forms of telling (1983, p. 222). The Boston

Table 7.1. The therapist's processing of communication.

Patient	Memory buffer	Stimulus type	Clinical function
Speech	Auditory buffer	Verbal	Primary
Text	Visuospatial	Visual/verbal	Infrequent
Body movements	Visuospatial	Nonverbal	Sensory
Gestures	Visuospatial/ kinetic	Metaphoric*	Visuomotor
Tone of voice	Auditory	Nonverbal	Emotions
Facial expressions*	Visuospatial	Nonverbal	Emotions**

Sources: *Cienki & Müller (2008). **Cozolino (2002).

Change Process Study Group concludes that much of the psychotherapy process depends on cues that are nonverbally based. Their focus is particularly on facial expressions as having "inherent meaning as positive or negative communications in and of themselves" (2010, p. 129).

Other nonverbal means of communicating, such as gestures and pauses, have been identified, but it is striking how little of their own nonverbal processing therapists describe in their reports. Gestures, for instance, often convey a metaphoric meaning and thus represent embodied cognitions, even though this meaning may be culturally determined (Cienki & Müller, 2008).[2] Pauses and silences, while making both therapist and patient acutely aware of their internal dialogues, also focus the therapist's attention on body movements and facial expressions.

One of the reasons for nonverbal communication's being unnoticed, as LeDoux (2002) points out, is that the processing of what is being held in the slave systems goes on unconsciously. By the time the therapist makes notes on a session, what he or she consciously experiences is the result of processing verbal information—other forms of communication do not remain available to consciousness unless they are of the most unexpected and attention-grabbing sort.

LeDoux describes a hypothetical scenario of how our attention changes from reading a newspaper when we unexpectedly hear someone mention our name.

> Although your conscious mind was ignoring everything apart from the visual signals from the paper, your brain was not. Inputs

from other sensory systems continued to be actively processed, otherwise the mention of your name could not have interrupted you.

(2002, p. 191)

In this example, someone mentioning our name has certainly captured our attention. Perhaps there was also a momentary lapse in our interest in the newspaper. However, if the other person in this example asks, "What's that smell?" our attention is quite likely to be shifted away from the newspaper and we "would probably start sniffing in an effort to answer the question" (ibid.). The shifting of mental resources—and with it the necessary behavioural changes—is something that happens apart from consciousness: in this example, from visual to verbal, from verbal to auditory, and from auditory to olfactory. And only the information that found our immediate attention is maintained for long-term storage and future use.

Where the executive reigns

Since the slave systems hold information only for seconds before it decays, the material held in them must be processed quickly. We must rely on our abilities to constantly compare, contrast, and predict what the buffered information means. These abilities reflect what the researchers call *executive functions*, a limited-capacity attentional system. The ability to integrate information from and across the slave systems allows for abstract representations of objects and events. It also allows us to consult what has already been stored in long-term memory.

To LeDoux, the executive functions are similar to a computer's operating system:

> In complex tasks involving multiple kinds of mental activities, executive functions plan the sequence of mental steps and schedule the participation of the different activities, switching the focus of attention between activities as needed. Executive functions are crucially involved in decision-making, allowing you to choose between different courses of action given what is happening in the present, what you know about such situations, and what you can expect if you do different things in this particular situation. Executive functions, in short, make practical thinking and reasoning possible.

(2002, p. 178)

Most researchers locate these functions in neural circuits in the frontal lobes, a large convergence zone that accounts for about one-third of the mass of the human brain. Located in front of the movement-control regions, the frontal lobes are especially developed in primates. Information from various specialised systems, such as the visual and auditory systems, serves a critical role in orientation and everyday survival.

But the executive also has connections to the hippocampus and other cortical areas involved in long-term explicit memory. The executive is therefore able to tap into knowledge from past experiences. LeDoux describes these pathways in relation to visual information:

> Thus, the pathways from the specialized visual areas tell the prefrontal cortex "what is out there" and "where" it is located. Moreover, these are two-way streets: the prefrontal cortex, by way of synaptic pathways back to the visual areas, instructs the visual areas to attend to and stay focused on those objects and spacial locations that are being processed in working memory.

> (2002, p. 182)

Similar linkage also appears to occur when it comes to auditory processing, and the two-way direction LeDoux describes is another name for the *bottom-up* and *top-down processing* mentioned above. The bottom-up information is supplied by the various posterior areas of the brain, such as the parietal cortex for verbal material, while top-down processes are yet another name for executive functions, the "top" being the prefrontal cortex (p. 183).

Working memory can therefore be defined as "what we are currently thinking about or paying attention to" (LeDoux, 2002, p. 176). Researchers examining damage to the prefrontal cortex found that deficits in attention and problem solving appear to be the main results of such damage (Eichenbaum, 2002). Thus, attention and what we train ourselves to pay attention to play major roles in how long-term memory functions are put to use.

For therapists, in particular, executive functions are geared toward what is being activated from past sessions with a patient—what I have called *session-related memory*. As I described in more detail in Chapter Four, memory for this information can clearly be identified as episodic, and it is one of the conscious, or explicit, memory systems. As Daniel Schacter of Harvard University notes:

In order to be experienced as a memory, the retrieved information must be recollected in the context of a particular time and place and with some reference to oneself as a participant in the episode. The psychologist Endel Tulving has argued that this kind of remembering depends on a special system called *episodic memory*, which allows us explicitly to recall the personal incidents that uniquely define our lives.

(1996, p. 17)

Session-related memory fits this description. The temporal or spatial cues necessary for its retrieval are provided by the ritual of psychotherapy. Sessions are generally conducted for a specific duration and in the same place, usually at the same time, on a regular, often weekly schedule.

Of course, this does not mean that we remember everything—or even most things—about a particular session. As we have seen, memory traces weaken shortly after each session unless they are consolidated and rehearsed via reporting or note taking. A large portion will, in fact, only be held in working memory for seconds and is then lost permanently. By the time that the therapist wishes to report on a case, these memory traces have become condensed and transformed and may only be available through certain cues or via narratives (Strupp & Binder, 1984).[3]

According to Tulving, a certain state of consciousness is required that "includes the rememberer's belief that the memory is a more or less true replica of the original event, even if only a fragmented and hazy one, as well as the belief that the event is part of his own past" (1983, p. 127). Only what was firmly encoded when the event occurred will, therefore, remain stored to be retrieved in future sessions.

The role of cueing

In order to access this information at a later date and in future sessions, therapists must rely on their memory being cued. And since episodic memory is primarily cued contextually, when the externals such as time and space are repeated, therapists' executive functions are often geared toward what they retrieve nonverbally. Thus, the cues are not limited to what was encoded from previous verbal exchanges. As we saw in Chapter Four, access to a unique treatment narrative for each patient is possible as long as the physical space and the emotional tone in which the initial encoding occurred are repeated. Once a skeleton of this narrative is established, the therapist's attention, aside from regular process notes, is mostly informed by such nonverbal and contextual cues.

This may explain why the contextual cueing and retrieval of episodic memory plays a significant role in treatments, both for the therapist and for the patient. When communicating over the telephone, most of these cues are missing, although we quickly adjust to having only sounds to go by. The cueing and retrieval of episodic memory during telephonic communication is therefore less likely; only face-to-face interaction can guarantee it.

This does not mean that semantic knowledge has no role to play (see Table 7.2, below). Certain facts and certain propositional knowledge form the background for assessments that the therapist makes throughout a treatment. Procedural and implicit functions are also at work; they are based on general attitudes toward practice that the therapist learns through repetition. What is significant about the therapist's use of memory, however, is how it remains dependent on the contextual retrieval of what is held narratively. This is how the therapist is able to have an understanding about what has occurred up to a certain point in a given therapy and be a full participant in what happens in new sessions.

For the patient, cueing of episodic memory may initially be of secondary importance. Patients generally enter therapy with a range of prepared questions and concerns about why their lives feel empty or their relationships are dysfunctional. Only when a therapeutic alliance has been established and a level of trust in the therapist achieved do they begin to use the sessions as a place where their autobiographical knowledge can be accessed and a new self-narrative created (see also

Table 7.2. The therapist's memory systems.

Memory system	Buffer	Cues	Type
Semantic data	Auditory	Verbal	Explicit
Episodic data	Auditory/visuospatial	Contextual	Explicit
Procedural data	Visuospatial	Nonverbal	Implicit
Emotional data	Facial/bodily	Nonverbal	Implicit

Chapters Three and Four). Episodic memory is, therefore, the patient's direct link to knowledge of his or her personal history. This knowledge, in turn, will influence how the patient experiences current situations.

In an instance of such retrieval, a sixty-year-old patient suffering from a longstanding depression began the session by describing new bouts of self-incriminations. He told the therapist how he is incapable of relating to others, how he thinks he has permanent memory loss, and how he has been cursed by illness throughout his life. The therapist quickly realised that this was a relapse after several years of therapy. At first, he focused on the patient's change of mood and explored possible situations in which the current difficulties originally surfaced. While listening to his patient tell about events that had occurred since the last session a week earlier, he was also cued by the patient's mood. The therapist remembered an earlier session in which the patient recalled his father's withdrawal and lack of interest in him. When the therapist reminded the patient about his struggles as a teenager and his memories of his father, in particular his memory of his father sitting in his favourite armchair for hours without speaking, the patient recalled how the current bout of depression started. An errand had taken him to the neighbourhood of his high school a few days prior. High school, he remembered, was when he felt estranged from others and deeply depressed.

This recall could not have happened without the therapist's having trained his executive functions to be tuned to nonverbal cues and accessing autonoetic awareness. Without any doubt, such cueing of episodic memory often plays a critical role in the therapist's ability to identify the patient's emotional state, and the retrieval is highly dependent on context. The physical space of the consulting room and the emotional tone that is set during sessions are, therefore, tied to the initial encoding and to previous disclosures in the therapy.

The prefrontal cortex

An important reason for the close connection between episodic memory and executive functions is their location neuroanatomically. Executive functions are distributed throughout wide areas of the prefrontal cortex; these areas appear to be critical in the retrieval of long-term memories, especially those associated with episodic memory. During retrieval, these memory traces are transferred from cortical storage circuits and into working memory (LeDoux, 2002).

As we discussed in Chapter Two, areas of prefrontal cortex also appear involved in the encoding of episodic memory (Wagner, Maril, & Schacter, 2000). Thus, for episodic memory to be encoded in the first place, the information has to be present in working memory—in other words, becoming conscious of one's past requires that the material *initially* be available in working memory. For retrieval, the material *again* has to be brought into working memory. This all seems to occur in the same or similar prefrontal locations.

In many models for how psychotherapy works, access to past experiences, especially those resulting in trauma and remaining unintegrated by consciousness, is regarded as central to the understanding of what cures (Cozolino, 2002; Schore, 2003; Siegel, 1999). Research on working memory appears to confirm this understanding and places a patient's here-and-now interactions with the therapist as the major agent for the patient's retrieval of past experiences previously dissociated from consciousness. However, the psychotherapy literature seldom describes the role played by cueing, in particular the therapist's cueing of what has already been communicated by the patient, often nonverbally.

The role of nonverbal cues is perhaps the most underestimated aspect of the psychotherapy process. It is here that the unique conditions in which psychotherapy is conducted play a critical role. How to document all the communication—and not merely verbal aspects or the therapist's narrative reported after the fact—is therefore a major challenge. The current reporting formats certainly do not allow room to account for any of the ongoing and nonverbal cueing. And without accounting for episodic memory as the structural element in the therapist's involvement, the reporting becomes very incomplete and often fictional.

Donald Spence of Rutgers University, in his groundbreaking examination of psychoanalytic reporting, identified this phenomenon:

> Just as we use our analytic competence to find the themes beneath the surface, so in looking back at a finished session we may easily transform the uneven, interrupted, shifting journey over a range of shifting terrains into a finished episode with one underlying theme. We may remember the beginning and end of an episode and unconsciously fill in the middle, even in cases where there was no middle and either patient or therapist failed to make the transition.

(1982, p. 247)

The therapist's memory of a session may, to Spence, consist of selected perceptions of the patient's nonverbal communication as well as of the latter's internal process during the dialogue. Spence calls these elements *privileged competence*, in that the information "belongs to the analyst at a specific time and place in a particular analysis" (p. 216). As such, this information is always incomplete. Spence continues:

> No matter how complete our knowledge of the treating analyst, it will never transform us into the same "analytic instrument" and there are many aspects of the analyst's privileged competence that, unless they are added to the text, we will never learn about in the first place.

> (p. 217)

Based on narrative analysis and without mentioning the role of cueing, Spence explains why all case reports suffer from a lack of semblance to what goes on in a therapy session. His observations confirm what many therapists before him had arrived at (see Chapter Eight). Freud, who in many ways created the format for case reporting, also noticed that his reports became "botched reproductions"; this in spite of his careful process notes, often recorded at night after that day's sessions (June 30, 1909, p. 238).[4] The therapeutic narrative only accounts for a very limited sense of what occurred in a given session.

An encounter barely remembered

In this perspective, therapists use other—sometimes several other—types of memory systems to report on their work (see Table 7.2). Since some of the most critical information from a session can only be retrieved in its original context, and with some reference to themselves as participants, therapists have an easier time with what was encoded semantically; that is, with certain facts and propositions that are part of a psychological assessment.

This is where the typical case report begins; I will refer to this type of memory as *general clinical memory*. Like all professionals who treat, counsel, or advise, therapists use this information to establish a profile of each patient and to generate possible approaches to patients' problems. A semantic system of factual information about the patient—such as age, parentage, siblings, occupation, marital status, and place of birth—general clinical memory results in an expository script that often includes a diagnosis.

The literature on psychotherapy research offers plenty of examples of this type of memory, in particular from initial interviews and clinical assessments. In their book *Psychotherapy in a New Key* (1984), Hans Strupp and Jeffrey Binder, psychotherapy researchers at Vanderbilt University, describe several cases using a particular form of short-term therapy. In one of those case descriptions, they summarise what appears to be the first session with a patient. They write:

> Arnold was a bright high school graduate in his later twenties, who complained of depression and anxiety resulting from "inability to achieve goals." When questioned further, he stated that although he had "always wanted" a higher education, he found himself unable to take any steps to realize this goal, despite ample financial resources and the apparent support of his family and friends. Arnold sought psychological consultation to determine whether his education aspirations were realistic and to help him get unblocked.

> (p. 82)

As is standard in these reports, the information comes from an explicit verbal dialogue in which the clinician asked the questions and the patient supplied the answers; it is the clinician who now reports what he asked and how the patient answered. Since there are some direct quotes, we have to assume that the reporting is based on notes taken during the session or shortly afterwards.

This procedure, originating in the medical examination, allows the therapist to introduce the patient's basic information without revealing his or her identity (Eells, 2007). In turn, the therapist uses this information to determine a diagnosis and to formulate a treatment plan, much like a psychiatric assessment. For the purpose of a psychotherapy treatment, however, the format has some obvious shortcomings. While the focus in a medical examination is primarily on the patient's physical status, reporting about psychotherapy involves many other elements, such as the patient's relational and social life, education, work life, cultural and religious background, and psychological history—all elements whose reporting is based on interpretations, not mere factual observation.

Joseph Schwartz, in his book on the history of psychoanalysis (1999), finds the field's dependence on medicine and psychiatry—the so-called medical model—unfortunate. For over fifty years, the field's insistence on medical training for would-be psychoanalysts limited the background of those practising in the United States. When the psychiatric monopoly was finally lifted, psychotherapy practice also became more heterogeneous.

> Psychoanalysis never should have become a medical specialty. But in the polarized atmosphere that has accompanied the medical disenchantment with psychoanalysis, what has been lost yet again are opportunities to advance our understanding of the interrelationships between the somatic and the psychological.

> (1999, p. 278)

General clinical memory, as reflected in case descriptions like Strupp and Binder's, today serves the purpose of establishing the overall status of the patient when he or she entered treatment. Depending on the clinician's background and education, this type of memory may vary considerably, but basically it is what allows the therapist to develop specific memory for what happens in each session.

The technique issue

General clinical memory is assisted by a certain type of implicit and procedural memory that therapists develop as part of practising. I call it *technical memory*, as it consists of certain habitual attitudes and abilities that are technical and procedural: greeting the patient, arranging seating, timing a session, and—as in the case above—conducting an initial assessment following a set of exploratory questions. Even some of the seemingly personal responses to a patient, such as dealing with questions and showing concern, are, to a large extent, based on technical memory. It is something the therapist does without having to think.

All such attitudes and abilities must be regarded as technical or procedural even when, after the fact, their resultant actions are thought of as "consciously executed techniques". Due to the implicitness of technical memory, therapists rarely think about how these procedures develop.

However, since this memory consists of habitual attitudes brought to the entire practice, it is sometimes included in descriptions of a particular personal approach. The paper by D. W. Winnicott, referred to in the previous chapter, has several of these references. According to Winnicott, the analyst functions in "a special state, that is, *his attitude is professional*" (1960, p. 161; emphasis in original). He continues:

> The work is done in a professional setting. In this setting we assume a freedom of the analyst from personality and character disorder of such a kind or degree that the professional relationship cannot be maintained, or can only be maintained at great cost involving excessive defences.
>
> (ibid.)

What Winnicott discusses as "a special state" must be retained in procedural memory; this special state is therefore associated with technique and professionalism. Similarly, Nancy McWilliams, in her book on psychoanalytic case formulations, describes her way of practising:

> Thus, unlike some clinics in which there is an intake process separate from a psychotherapy referral, in my practice, the intake session is usually the beginning of the ongoing relationship between a patient and me. Most of the people who come to me are voluntary and self-referred, and although this group contains a fair number of individuals with borderline and psychotic psychopathologies, few of the prospective clients who appear at my door are frighteningly disorganized or dangerous, or in need of immediate hospitalization.
>
> (1999, pp. 30–31)

This type of practice is particularly common among psychoanalysts. Downplaying the role of standard intake procedures may, in fact, be a privilege for clinicians in private practice with a very specific patient population. To McWilliams, technical memory thus boils down to an assessment of "the person's reactions to whatever notions I have about how to make sense of the problems described" (1999, p. 31). Naturally, her patients' reactions must be based on things that McWilliams does habitually, without much forethought.

Interpretative memory

As Strupp and Binder continue to describe the case of Arnold, they no longer focus on general information. In exploring the patient's complaints further, the therapist in the report observes that the patient was "overly chummy". As an example, he mentions that in the first session "Arnold addressed the therapist by his first name and behaved as if he and the therapist were old friends" (1984, p. 82). This observation leads to the following hypothesis:

> The therapist suspected that this behavior might be an effort to prevent an opposite (warded-off) kind of relationship. He hypothesized that, by behaving in a gregarious manner, Arnold was attempting to test (and perhaps foreclose) the possibility that the therapist might not like or accept him.

> (p. 82)

This hypothesised explanation originates in a script that the therapist developed well before beginning the therapy, now being activated by what Arnold says but also by what he does not say. In other words, the therapist's script has been cued by Arnold's nonverbal expressions, possibly his tone of voice or gestures. To the therapist, Arnold's behaviour is a defence against rejection based on an "offence is the best defence" strategy. The script may therefore be as follows: *By being expressly friendly, a person may seek to convince others not to be rejecting, at the same time avoid feeling vulnerable.*[5] Translated into a proposition: *When a patient is overly chummy in the initial phases of a treatment, it may be a sign that he or she fears rejection, thus wishes to be treated as a friend.*

We have called these narratives *therapeutic scripts* (see Chapter Four). They consist of propositions, predictions, and beliefs—hypotheses that therapists nevertheless rely on, in particular in initial phases of a treatment and before a full treatment narrative has emerged. Therapeutic scripts are therefore preexisting long-term memory traces. Generally, therapists are not aware of when and how they acquire them. As discussed below, they are, nevertheless, tied to therapists' paradigmatic formulations and certain concepts attached to them.

Not all scripts are propositional. Some are based on recommendations that may go back to the therapist's training and a supervisor's suggestions. Some may be in the form of predictions, and still others are simply beliefs that the therapist would find difficult to verify. Together, however, the scripts form a type of professional expertise that therapists acquire through experience.

What scripts have in common is that they give the therapist a sense of what a disclosure (verbal or nonverbal) may mean—what developmental psychologist Robyn Fivush describes as a generic memory of "what happens each and every time the event occurs" (1997, p. 142). Most scripts remain internal references in the therapist's mind until they can be confirmed by explicit patient disclosures. There is no doubt, however, that they influence the therapist's interpretations of what is being disclosed in a session. And on occasion, they manifest in direct interpretations.

In the case of Arnold, we are in the unique position of knowing the source of the therapist's script. Strupp and Binder are also the authors of *TLDP Focal Narrative* (1984), which lays out a specific psychodynamic approach to short-term therapy that involves four categories: acts of self, expectations about others' reactions, acts of others toward self, and acts of self toward self (introject). The authors explain:

> Like many other conceptual tools available to a psychotherapist, the TLDP focus format is primarily a heuristic aid for the clinician and should not be mistaken for what is ordinarily communicated directly to the patient.

(p. 77)

In introducing their focal narrative, the authors also confirm that what we are dealing with is not a paradigm or school of thought but a generic structure for narrative understanding. To them, "[T]he content of a TLDP focus may be drawn from any number of broad story metaphors, cultural myths, and psychological theories of personality" (p. 78). The TLDP focal narrative is, in fact, a tool—a system for organising a collection of scripts.

In their case vignette about Arnold, Strupp and Binder confirm that the script initially was an internal reference for the therapist. Only after direct verbal disclosures by the patient was the meaning of the script revealed.

> While this pattern of seeking acceptance was first noticed in the therapeutic relationship, the therapist soon discovered other

interpersonal contexts in which to elaborate this observation and
relate it to the present complaint.

(1984, p. 82)

The script is now accounted for as an "observation", while it seems
more likely that it was via nonverbal cues that the therapist formed
the initial hypothesis. The clinician's confirmation of this hypothesis in
other interpersonal contexts is a subsequent development.

The treatment narrative

The sequence involved, from scripts to specific observations, is a common one in how therapists process their perceptions of a patient. Initially, the therapist's memory of specific verbal disclosures in a session
will consist of select perceptions of the patient's nonverbal communication as well as the internal scripts activated during the dialogue. As
a specific transactional pattern emerges, however, these more impres-
sionistic elements are replaced with specific episodic memories, and the
therapist can now rely on session-related memory. (See also Table 7.2,
above.)

As such, this information is always incomplete. To Strupp and
Bender, this means that a patient's problems have to be understood in
transactional terms. The therapist's role, in fact, is to seek transaction
and to evaluate what they call "the functional salience of identified
transactions" (1984, p. 80). And in order to do so, the therapist has to
retrieve enough relevant episodic memory particular to each treatment.
This also assumes that both therapist and patient have begun to form
narratives tailored to fit the unique circumstance of working together—
in other words, a working alliance. Although by necessity different,
these narratives are now each participant's record of the treatment, in
narrative form.

This record is also based on narratives that are dynamic and evolving. For the therapist, the narrative is a treatment narrative that gradually transforms into an observer story. Through this story, regressions,
impasses, and repairs in the therapy will be remembered as part of a
process. So will turning points, resolutions, and integrations. Organised
with a distinct set of characters, usually reflecting the patient's original
family constellation and relevant experiences in the past, the treatment

narrative is what primarily guides each of the therapist's interactions with the patient.

In the process, the first-person nature of this narrative will change. The therapist is no longer the story's protagonist—as many of the episodic memory traces become inert, the therapist instead becomes the story's narrator and observer. After all, the therapeutic relationship functions in the privacy of a confidential relationship, with very few social implications, and is limited by the treatment ritual of weekly sessions. The patient's integration and healing are the purpose of the treatment.

The patient's narrative, on the other hand, must go through many revisions and condensations and will be experienced as a series of mini-stories that eventually are combined into an overall story about the treatment. In the end, it becomes a story directly linked to the patient's autobiographical self and first-person knowledge.

Paradigms revisited

For the therapist, the treatment narrative is also tied to another type of memory, which I will call *paradigmatic memory*—a set of master narratives maintained to reflect a certain paradigmatic approach. These narratives are usually stable over time and allow therapists to formulate hypotheses about their work, make theoretical statements, and present how they work with patients. In this respect, paradigmatic memory serves as an overarching narrative for many of the therapist's scripts.

Although of a narrative origin, these scripts are expository and lack the typical storylines associated with narratives about personal experiences. Instead, they consist of frames, or models, that therapists were taught during training. As concepts and definitions—categories that are held in semantic memory—they explain the patient's symptoms and personality but, due to their semantic nature, they are unable to hold information about what occurs with a patient from session to session.

As shown in the previous chapter, occasionally therapists outgrow the old paradigms and develop new theoretical bases. This was the case for Heinz Kohut (1973), the Chicago psychoanalyst who formulated the self-psychology paradigm. Well into his career, his clinical experiences came in conflict with the traditional analytic paradigm in which he received his training. Others, such as Paul Wachtel (1992), started out as classical psychoanalysts, later to embrace more of a cognitive-behavioural approach.

Although theoretical formulations clearly originate in the literature, they most likely find their particular personal meaning when the therapists first try to understand themselves; thus, they go back to their personal analyses (Levine, 1994). But do they remain the same after the therapist has practised for several years? All indications are that they change and become highly individualised with time (Fabricius, 1995). The terms and definitions may stay the same, but as dynamically evolving narratives, theoretical formulations become the practitioner's special version of Freud, Jung, Klein, and so on.

The question, then, is why theoretical formulations are not abandoned altogether. They certainly appear rather insignificant when therapists are processing immediate experiences with patients (Cambray, 2001). Well-encoded through rehearsal, theoretical formulations resurface when therapists seek to legitimise each patient involvement. They are the concepts and storylines necessary in giving meaning to an entire process.

As stories, however, theoretical formulations are immune to scientific testing. A better name perhaps, is *skeleton stories*, as Roger Schank (1990) of Northwestern University calls them, since they are generic constructs that no longer can be traced to specific experiences. However, as templates they will determine the meaning of the experiences to be described. They lend credibility to the therapist's work when it is presented to the public or to colleagues. Rather than capturing interactions with a patient, theoretical formulations thus tend to leave out the therapist's own involvement while offering predetermined explanations for the patient's pain and afflictions.

Therein also lies the danger of these broadly based paradigmatic formulations. They may appear to be objective descriptions of particular procedures used with a particular patient, but they are constructed around certain narrative formulas. Their purpose is to persuade more than to be exacting accounts of what happened in therapy (Hillman, 1983).

Freud and other master stories

The psychoanalyst Roy Schafer was one of the first to propose that therapeutic theories, in reality, are what he calls *master narratives*: broad interpretative structures that serve as an overall guide to the therapist's understanding (1980). Freud's theories, he found, translate into two such narratives: one of the beast, the other of the machine. The beast story, according to Schafer, tells how the infant, also called the *id*, starts out as a beast and must endure the frustrations of being tamed into a civilised world, away from nature. And even though the taming leaves each person with two regulatory psychic structures, the *ego* and the *superego*, the protagonist remains, in part, a beast and the carrier of an indestructible id.

In the machine story, on the other hand, Schafer recognises Freud's metapsychology, a vision of human nature seen through the lens of the physiological and neuroanatomical laboratories of the nineteenth century (1980). According to this tale, the mind is a closed system, an inert mental apparatus that needs the force of instincts to function. Ruled by Newtonian physics, its amount of energy is fixed: when some of its energy is expended, less is available for other purposes.

Schafer argues these archetypal stories have been "mythologically enshrined" (1980, p. 28). The fact that other psychoanalytic variants have developed over the years proves that we are dealing with narratives, not scientific theories.

> That Freud's beast and machine are indeed narrative structures and are not dictated by the data is shown by the fact that other psychoanalysts have developed their own accounts each with more or less a different beginning, course and ending.
>
> (p. 29)

A similar reading of Jung would tell of a child's secret connection to God, revealed in dreams and visions in search of eternal life. Only through what he calls *individuation* will this secret connection be maintained and humankind freed from its destructive loss of meaning (Jung, 1935b).

Yet other narrative structures than those mentioned by Schafer can be read into the writings of Freud or Jung. Kohut, for one, read Freud as describing what he calls Guilty Man, a Victorian creature ruled by repression and Oedipal conflicts; and James Hillman (1983) has shown us several ways of reading Jung, including as a tale of obsessive mono-theism in which all development is geared to the Self, with a capital *S*.

Since Schafer's paper, narratives have become a popular way to explain a variety of phenomena in analysis. Most writers, if not all, have used it to justify established theory. Marshall Edelson (1992) illustrates how narrative meaning can be attached to patients' fantasies using traditional Freudian theory; Coline Covington (1995) details the role of narratives in analytic interpretations from a Jungian/Kleinian perspective. Robert Winer (1994) looks at case reporting as a type of narrative fiction, possibly inspired by Hillman's early explorations of the topic.

Winer also hints at the possibility that we are discussing dynamically evolving memory structures of a highly individualised nature, and that each practitioner creates and recreates his or her master story. From that vantage point, there are certainly many other master narratives in use, with different plots and different characters. Some may be combinations of the ones already mentioned, others may stake out entirely new narrative grounds.

The terms used in these stories are far less significant than the fact that therapists develop them for a particular purpose. Even when a very similar set of terms is used—such as *ego, psyche, the unconscious, transference*, and *countertransference*—the terms themselves tell us very little about the particular events attached to them. This problem has only recently been acknowledged by the therapeutic community. Paradigmatic theories clearly play a very secondary role in treatments, so why are they still in use and argued over? Peter Fonagy, of University College–London, concludes that they only have meaning for the therapist:

> But do theories matter at all? Do they really influence clinical work with patients? This is a difficult question to answer. Evidently, analysts from very different persuasions, with very different views of pathogenesis, are convinced of the correctness of their formulations and are guided in their treatments by convictions. Since we do not yet know what is truly mutative about psychotherapy, it might well be that for many patients the analyst's theory of their etiology is not so crucial.

> (2005, p. 131)

As shown previously, many psychotherapy researchers, among them Jerome and Julia Frank (1991), have reached this conclusion as well.

CHAPTER EIGHT

What there is to tell

E ver since Freud abandoned his attempts to anchor psychological principles in neurobiology, the psychotherapeutic field has shown an unfortunate lack of concern for the role memory plays in psychotherapy treatments. What, and how, patient and therapist remember has received little attention in spite of the rich source of data from neuroscientific research that is now available. In fact, today, in examining what can reliably be reported about a specific treatment, we must take into consideration how memory functions.

By the time Freud issued his technical recommendations on how to conduct treatments, he had abandoned his ambitious "Project for a Scientific Psychology" and was in the middle of developing his psychoanalytic methods (1885).[1] He now based his formulations on what he called the *case history method*, a narrative rather than a neurocognitive approach (Gay, 1998; Messer & Wolitzky, 2007). His technical recommendations nevertheless give us a fairly clear picture of his own approach to remembering what is significant in each session and about each patient. The technique, he writes, "is a very simple one" and involves the following:

> As we shall see, it rejects the use of any special expedient (even that of taking notes). It consists simply in not directing one's notice to anything in particular and in maintaining the same "evenly-suspended attention" (as I have called it).

(1912e, p. 111)

The purpose of this quiet attentiveness, he explains, is to avoid focusing on what is already known or falsifying the information in some way. Instead, the therapist should "give himself over completely to his 'unconscious memory'" (p. 112). Freud does not elaborate on what the therapist's unconscious memory consists of, but it is clear that he envisioned it to be cumulative. Accordingly, what the therapist holds in memory would become conscious "as soon as the patient brings up something new to which it can be related and by which it can be continued" (p. 112). Freud, in other words, relied on his episodic memory and its contextual cues in order to remember previous disclosures by his patients.

Case formulations today

The literature about case formulations today covers a wide range of theoretical orientations to psychotherapy, but many of Freud's practices are still in use. One of these practices has to do with note taking. Freud discouraged note taking during sessions and preferred to write down examples "from memory in the evening after work is over" (1912e, p. 113). Such note taking, still used in current therapies, was in fact the basis for his case reports, a type of progress notes that continues to be a critical aid for clinicians. When considering how memory functions, note taking from the same day still allows a good amount of trace material to be accessible and many of the therapist's own contributions to the therapeutic dialogue are still possible to retrieve.[2]

Note taking during the session is another matter. As we saw in the previous chapter, research on working memory and attention suggests that, for the processing of nonverbal information, concurrent note taking may limit the therapist's ability to process such information. In addition, when one is processing auditory information, one's attention is best focused on as few tasks as possible.

The most recent *Handbook of Psychotherapy Case Formulation* (Eells, 2007) includes a traditional psychoanalytic perspective on how to make such formulations. The handbook also includes contributions from several psychotherapy researchers as well as chapters on cognitive-behavioural and dialectical-behavioural therapies. Still, none of these authors mentions the role of memory in psychotherapy. This is remarkable considering that the formulations they discuss are all made after the fact—months after the treatment began—and thus rely on what the therapist remembered. Any formulation, even those made within twenty-four hours of the sessions, can no longer be regarded as complete and free of distortions, especially since we know that the retrieval of episodic memory depends on the consulting room as the time and place context.

Tracy Eells, the handbook's editor, describes case formulations as critical tools in planning and anticipating what a treatment will be all about. They provide "a link between theories of psychotherapy and psychopathology, on the one hand, and the application of these theories to a specific individual, on the other" (2007, p. 25). What Eells does not discuss are the different formats in which these formulations appear, and that many of them are narrative. For instance, the most obvious format for case formulations is the published report, used in self-help books as well as in professional journals. Other equally important uses are in the training and education of future psychotherapists (Messer & Wolitzky, 2007).

A closer look at self-help and professional journals reveals few fully developed case formulations. Rather, in both instances, short case vignettes are mostly what comprise the reporting, be it for the general public or the professional community.[3] The same is true in the training of future therapists. These reports, for reasons of confidentiality, are usually not available beyond the institutions in which they are produced. In all these instances, many of the identifying details have been altered and the material is mainly anecdotal. The accounts are exclusively based on the therapist's narrative and cannot be checked for accuracy.

When Freud introduced written case histories as narratives of a specific treatment, he usually described completed cases. Today, the term *case formulations* mostly refers to treatment plans formulated at the beginning of a therapy. These plans follow a common, standardised format. Accordingly, Eells defines case reporting as "a hypothesis about the causes, precipitants, and maintaining influences of a person's psychological, interpersonal, and behavioral problems" (2007, p. 4). What this definition does not address, however, is how case formulations will be used beyond initial assessments and how they are modified by the treatment process. Historically, case reports covering entire treatments have had—and continue to have—a significant influence on how psychotherapy is practised. How this type of reporting is to be done today is less clear in the current literature. This omission is particularly relevant, since case material is a critical component in the teaching of psychotherapy skills.

Also, the therapist's hypothetical understanding is only one of several factors influencing a treatment and is bound to undergo many changes. As Sheila Woody and colleagues conclude in their book on treatment planning, selecting methods that have support in psychotherapy research is only one of many possible steps (Woody, Detweiler-Bedell, Teachman, & O'Hearn, 2003). Treatment planning involves how the therapist "figures out the specifics of the treatments and how to conduct them" (p. 4). And even with a good treatment plan, a successful therapy will be influenced by what occurs throughout it, not least because of the ways the patient responds to the therapist's plan (Binder, 2004).

In fact, research and practice have different and incompatible goals. In psychotherapies, documentation and reporting are used to produce a record and describe certain approaches, certain attitudes that are part of a treatment. The reporting highlights these approaches and attitudes without being able to do so extensively. In their reports, therapists share how they worked with a patient and discuss what they found effective in a particular case. Generally, they are not concerned with broader issues about the laws for how the data were collected.

Furthermore, therapists document probable antecedent experiences of their patients. Based on the patients' disclosures, therapists try to establish a credible narrative of the psychological and developmental history of each person they are seeing. This narrative, often continuing to develop as further antecedent experiences are revealed, will also become part of what therapists report. Although sometimes becoming a story-within-the-story, the narrative presentation of a patient's history is how therapists establish a credible record of their patients.

Change vs. causes

The obvious goal of all treatments, however, is to encourage change, not pinpoint its causes (Binder, 2004). Reliability and validity of data, as required in research, are beyond the resources at hand for the practising therapist. By documenting and reporting what they find to be true about a particular patient and a particular treatment, they try to convey how a successful outcome or failure came about. They do so by describing treatments as processes—with beginnings, impasses, or reversals, and, finally, terminations. Unless terminated prematurely, a treatment is experienced as reaching certain stations, such as establishing a therapeutic alliance, processing new revelations, and arriving at orderly completions (Ryle & Bennett, 1997). This is how patterns in human relationships are experienced, remembered, and communicated: in story form. Thus, therapists use narrative aids (McAdams, 1993, 2006).

This does not mean that the thornier questions about reliability (internal coherence) and validity (external verifications) are irrelevant. They often influence, in general terms, the type of approaches the practising therapist feels comfortable with. According to Eells (2007), a case formulation's reliability refers to how well other clinicians independently could construct similar formulations about the same clinical material. In other words, a case formulation is *only as good as the formulation that would replace it if the same data were given to another clinician*. Validity, on the other hand, refers to *how well a formulation predicts the outcome of a given therapy and its process* (ibid.).

Unfortunately, when it comes to describing entire cases, there is no way the data from one therapy could be given to another therapist in order to test the reliability of the first therapist's formulations. This is a luxury available only to psychotherapy researchers, who typically have expert judges listen in on earlier recorded sessions (Miller, Luborsky, Barber, & Docherty, 1993). To the practising therapist, the reporting will have to remain beyond these strict empirical standards (Woody, Detweiler-Bedell, Teachman, & O'Hearn, 2003). The only meaningful way the everyday therapist can describe his or her interactions with patients is as two-way communication: as the therapist's narratives and, more tentatively, as therapeutic narratives that include the narrative disclosures of patients (Wiener, 1994).

As to the validity requirement, we are on more solid ground with assessments and treatment plans than with full case reports. The former are formulated at the onset of a therapy and they often involve a comparison of the therapist's observations with the descriptive criteria in a diagnostic manual. In other words, as long as the focus is on symptoms and certain criteria for mental disorders, we have other descriptive sources with which an assessment or a treatment plan can be compared.

For a full case report, however—which attempts to describe developments within ongoing interactions between therapist and patient—there are no similar comparative sources. Every therapeutic relationship develops its own particular dynamics, its particular timing and pace (Margison & Bateman, 2006).[4]

Freud's dilemma

One of the common methods for aiding memory is keeping process notes. This practice may serve to counter the twenty-four-hour limit Conway (2002) discusses. Freud, for instance, made careful notes of his sessions each night, but, in a letter to Jung, June 30, 1909, he also complained that case descriptions based on his notes became "botched reproductions" (1909, p. 238). In case histories, he notes, "we pick apart these great art works of psychic nature" (ibid.). What Freud noticed about the difficulty of recalling past clinical events illustrates the context-sensitive nature of episodic memory. When a therapist tries to reconstruct what happened after a session and from written notes, many of his or her memory cues will be unavailable.

Freud made this observation about retrieval, but it did not stop him from writing his famous case histories. In spite of lacking certain episodic information, only available contextually, his case histories nevertheless present us with a flavour of what went on during his treatments (Freud, 1905e, 1909, 1911c, 1918).[5] Unfortunately, Freud's case histories have also left us with a template for reporting on cases that no longer meets important requirements. If significant traces of episodic memory cannot be retrieved outside the context of sessions with a patient, case histories will inevitably, using Freud's words, be "botched reproductions" and a great deal of what transpired will be missing. In their visual and emotional character, retrieved memories are therefore most suited to become shared references between patient and therapist.

Case formulations covering the process of psychotherapy certainly become less accurate as the therapist's memory becomes more and more inert. As we have seen, this is quite contrary to what Freud assumed. Seeking to validate his theories, he required a clear beginning and a clear end in reported cases. Unfortunately, after several years have elapsed, much of the therapist's recollection of what occurred is no longer available. Even if careful progress notes are consulted, the cases have lost important parts of their contexts.

Additionally, even if documented before much time at all has elapsed, as noted earlier, the reporting of what transpired between the two treatment participants is inevitably incomplete. What Freud (1912e) called "a synthetic process of thought only after the analysis is concluded" (p. 114) will only produce an outline and an approximation of the therapeutic sessions. To begin with, much of what occurred will soon be lost to, or never enter into, memory; and even when therapists use recording devices, important dimensions of nonverbal communication go unnoticed. Therapists' reports, by all accounts, remain condensed versions of what happened from their own perspective. The same is true, of course, when a patient describes his or her therapy.

The formats used for reporting, be they shorter case vignettes or longer reports on entire treatments, are nevertheless important. If nothing else, they help the reader to understand, with some degree of confidence, that what is described must be authentic. Among other things, this means that the reporting should include which stage of the treatment the material comes from, at what point in the treatment the report was written, and if the reporting was shared with the patient.

Research vs. *treatment*

In formulating his case history method, Freud made valiant attempts to go beyond showing that the reporting was authentic. Case formulations were meant to verify the specific and recommended approaches to treatment. Thus, he discourages collecting material for future reports while a case is ongoing since it would not achieve this goal. "One of the claims of psycho-analysis to distinction", he asserts, "is, no doubt, that in its execution research and treatment coincide" (1912e, p. 114). Ongoing treatments therefore had to be omitted, since the outcome was still uncertain. To verify that the patient entering therapy no longer suffered from a mental disorder upon its completion, the reporting would have to provide proof of the effectiveness of the psychoanalytic treatment (Spence, 1989).

In the same paper, however, Freud acknowledges that research and treatment are two different mental attitudes: "The distinction between the two attitudes would be meaningless if we already possessed all the knowledge (or at least the essential knowledge)" (1912e, p. 114). "At the present", he admits, "we are still far from that goal and we ought not cut ourselves off from the possibility of testing what we have already learnt and of extending our knowledge further" (pp. 114–115). In holding out hope for fuller knowledge of the unconscious, Freud nevertheless avoided a more fundamental question. If research and treatment are two separate attitudes, two separate approaches, how can both be used at the same time, in the same situation?

The future research in which he placed his faith has unfortunately established that the two approaches are fundamentally different and that research requires a level of control of variables that the therapy encounter does not permit. Research specific to psychotherapy can, under those circumstances, only be conducted on defined variables using large samples, tape-recorded sessions, and independent verifications (Miller et al., 1993). When the results from this research are standardised for general use, they have understandably met with considerable resistance among most clinicians (Woody, Detweiler-Bedell, Teachman, & O'Hearn, 2003).

Although Freud could not foresee that the gap between research and treatment would by no means be a temporary phenomenon, he also could not abandon his scientific aspirations. Thus, his case reports for a long time were given significance beyond their instructive value. They were mistakenly viewed as research reports rather than as pedagogic material that also happened to be good reading.

Case reporting to test established theories and learn more, on the other hand, continues to be an important tool. In order to test interpretations that feel too facile, note taking and reporting should be an integral part of any treatment. But when Freud proposed that future discoveries one day would replace ongoing learning about each patient, he negated his own statements about the importance of learning about each patient's history. The assumption that therapists one day would, instead, be able to rely on procedures and formulas has turned out to be unrealistic and erroneous. A patient's history and his or her past learning will continue to be a critical aspect of how the patient functions psychologically.

The important question for all therapists—and in all treatments—is how they use their memory to deepen their understanding of their patients and stay open to new treatment possibilities. As neuroscience arrives at new findings relevant to psychotherapy, these findings also need to be reflected in how the treatment process is understood so that the documentation and reporting of this process accord with how human memory functions.

A cacophony of theories

One of the reasons so little attention has been given to case reports in psychodynamic circles is because of our inheritance from Freud and the pioneers: an emphasis on paradigmatic correctness rather than descriptive criteria.[6] The problem with most paradigmatic formulations, as discussed in the first chapter, is their dependence on depth psychology, an old and, by now, outdated theoretical framework that in hindsight never really fit the clinical data (Ellenberger, 1970).[7] Donald Spence (1982), in examining the narrative underpinnings of psychoanalytic formulations, concludes that there never was a psychoanalytic theory, certainly not one that fit all the observations therapists make with their patients. He writes:

> It may be comforting to assume, as is the custom, that we have a general theory waiting in the wings, waiting to be confirmed; but the fact that after almost one hundred years we are still waiting for a set of confirmed postulates should give us a good grasp of future prospects. The widespread belief in this general theory may actually interfere with our clinical work.

(p. 293)

As Spence also makes clear, the clinical findings were "stored primarily as memories in the minds of practising analysts" (1993, p. 41). As long as this was a common store "in a heterogeneous set of minds", few of his colleagues noticed that their clinical information was "a shifting and unreliable data base which is not open to public inspection or consensual validation" (ibid.). Rather, the information was a closely held secret, a collective wisdom among those initiated into the practice of the psychoanalytic method.

What Spence concludes about the traditional Freudian group of practitioners can certainly be applied to the other analytic groups as well (Ekstrom, 2002a). As the communal sense fell apart—and it did so very early in these psychoanalytic communities—theoretical formulations appear to have become more, not less, important (Roazen, 1975; Schwartz, 1999). In presentations to the public or to colleagues, therapists would turn into defenders of the key concepts particular to the institute in which they trained, the therapist they trained with, and, most of all, the formulations of some prominent founder, whether it be Freud, Jung, Klein, or Kohut (Wiener, 1994). As a result, what we are now dealing with is a cacophony of theories, all claiming to explain the inner workings of patients' psyches (Ekstrom, 2002b).

Today, theories congruent with findings about memory must replace the depth psychological models in psychodynamic discussions of case material. To begin with, depth psychology was never a scientifically tested theory but a loosely organised philosophical argument by nineteenth-century thinkers (Ellenberger, 1970; Flanagan, 1989). Nevertheless, in the hands of the pioneers in the field—most of them trained in medicine—these ideas allowed methods for psychotherapy to be introduced. However, to an extent impossible when psychoanalysis was first developed, research data now available give a comprehensive framework for what occurs in a treatment. Based on neurobiology and neurocognitive models, they explain what both therapist and patient remember. This means a return to looking at brain function as the foundation upon which to postulate principles of psychology—ironically, the approach Freud attempted to follow in his early, preanalytic research (1895).

In hindsight, the first generations of psychodynamic therapists made the mistake of assuming that they could formulate a reliable theory to explain the causes for psychological disorders. In order to give credibility to what they discovered through working with their patients, they theorised in broad terms, always with an intrapsychic and one-person perspective of their patients. The result was a set of generalisations instead of a description of what happens between two people doing psychotherapy (Schafer, 1980).

What they left behind, as a consequence, is a language that, in the extreme, describes psychotherapy patients as inhabiting psychological bubbles, as if they were frozen in their pasts (Wachtel, 2008). On the one hand, these patients are supposed to be driven by an unconscious which only the therapist, as expert, has access to; on the other hand, the same patients are thought capable of the most outrageous distortions in how they view the therapist, the person with whom they are in dialogue. This language is descriptively weak, one-sided, and non-relational. At least implicitly, it is condescending toward the patient. It seldom describes any direct person-to-person interactions. Instead, it proposes to prove laws of general psychology based on very thin inductions (Spence, 1982).

The Dodo bird effect

Well into the 1980s, the psychoanalytic establishment continued to claim empirically verified methods and insisted that its theoretical and technical formulations should be viewed as science because its members framed them in biomedical terms (Kernberg, 1980). This was the justification for regarding psychoanalysis as a medical subspecialty; but research intended to verify its scientific status found no such substantiation (Shorter, 1997).

The design for these projects was commonly a variation on the medical model that says that specific ingredients, as presented in treatment manuals or psychodynamic formulations, are responsible for good outcomes. In a recent meta-analysis of such component studies, Hyun-nie Ahn and Bruce Wampold of the University of Wisconsin found this research futile. Their analysis, instead, supports what they call "the common factors model":

> The research evidence supports the notion that the benefits of counseling or psychotherapy are derived from common factors. For example, it has been shown that the therapeutic alliance, measured at an early stage, accounts for a significant portion of the variability in treatment outcomes.

> (2001, p. 255)

Similar views are expressed by Edward Teyber and Faith McClure, two psychotherapy researchers from California State University–San Bernadino, in the most recent *Handbook of Psychological Change*:

> Comparative studies of psychotherapy outcome consistently find that therapy modalities are relatively equivalent in effecting client

change. In contrast, there is considerable support for the view that
the individual therapist's attributes, attitudes, and actions (e.g.,
interpersonal skills, countertransference propensities, and personality) match or override the effect of particular techniques.

(2000, p. 80)

These conclusions, called "the Dodo bird effect" after a quote from *Alice
in Wonderland*, come after many years of trying to find the one approach
that is most effective (Wampold, 2001).[8] Fanciful theoretical formulations, as well as recommended techniques, may imply certain attitudes.
They do not capture what happens in a successful treatment, or, for that
matter, in an unsuccessful one (Ekstrom, 2002b).

The contextual model

In presenting the contextual model of psychotherapy, Bruce Wampold
and colleagues compare it with the traditional medical model, which,
in their description, assumes that sufficient psychological explanations
exist for all mental disorders. When the therapist administers ingredients derived from these explanations, change should follow (Wampold,
Ahn, & Coleman, 2001). While the medical model is based on the notion
that effective treatment must follow on a correct diagnosis, the contextual
model does not presume that mental disorders can be fully explained
by a diagnosis. The contextual model does not require a distinct diagnosis. Rather, it requires that a charged and confiding relationship can
be established within a healing context. Within this relationship, and
with the active participation of both parties, a therapeutic process transpires, as long as a rationale is present that gives a plausible explanation
for the patient's symptoms and is consistent with his or her worldview
(Wampold, 2001).

As a result of this shift, Ahn and Wampold (2001) assert that treatment protocols and other devices, such as mental status exams and
psychological assessments, lack research evidence. They may "appear
scientific and may be required for experimental control in the research
context", but, in their findings, they could also "cause ruptures in the
alliance, and consequently, poor outcome" (p. 255). Based on Jerome
Frank and Deborah Frank's suggestions in *Persuasion and Healing* (1991),
Wampold, Ahn, and Coleman (2001) instead advocate for a model
emphasising the following factors:

- an emotionally charged, confiding relationship with a helping person (i.e., the therapist)
- a therapeutic process that transpires in a healing context
- a rationale, conceptual scheme, or myth, that provides a plausible explanation for the patient's symptoms and is consistent with his/her worldview
- a procedure or ritual that is consistent with the rationale of the treatment and requires the active participation of both patient and therapist.

In this model, the patient will expect a relationship to develop in which he or she "divulges emotional and psychologically sensitive material" (Wampold, Ahn, & Coleman, 2001, p. 268). And as the therapist provides plausible explanations for symptoms, the patient also believes that the therapist will provide help and will work in his or her best interest.

Research attempting to demonstrate the efficacy of particular paradigms most likely will continue, even though the results so far give no support for such "special ingredients" (Strupp & Binder, 1984; Strupp, 1986; Teyber & McClure, 2000). As long as claims persist that one approach is superior to another, proponents of other equally valid approaches will have to prove the many benefits of theirs. And even if the contextual model proposed by Wampold and colleagues most adequately reflects what works in psychotherapy treatments, when it comes to initial assessments, the medical model and the DSM nomenclature have become standards unlikely to be abandoned (Chwalisz, 2001).

Beyond manuals and procedures

Beyond manuals for assessments and for identifying psychopathology, psychotherapists are left with many uncertainties, often looking for guidance about what approaches have a chance to work. In conducting entire treatments, psychotherapists are concerned not only with their patients' diagnoses and symptoms: of utmost concern are issues related to healing and how to facilitate deep and lasting psychological change.

Many of these issues are directly related to memory. As long as psychotherapy research focuses only on what method works with certain disorders, many aspects of treatments worthy of study will remain untested. Those variables related to memory deserve much fuller attention (Ahn & Wampold, 2001).[9] We still have very little information about how much of a session the average therapist remembers and how much of the remembered information is relevant to the patient. We also know very little about what memory skills are most relevant for the therapist, and about how best to train therapists, both to listen and perceive nonverbally and to hypothesise about what they hear their patients speak about.

This type of information would help therapists to develop credible records for what occurs in their therapies. Although some details will be unique to a particular treatment, other therapists may wish to discover the particulars for how this approach was helpful in a patient's process of healing.

When describing relational events

First and foremost, meaningful reporting needs to describe a therapeutic process by using language that reflects the relational nature of what happened between the two participants. This can be done in several ways, but since we are discussing narrative accounts, the reporting will, at least partially, depend on what is expected from the audience and the purpose of the reporting. This fact, after all, is the blessing but also the curse of narrative constructs. Depending on where and to whom the reporting is addressed, at least four different situations can be identified:

1. *When therapists use case vignettes to illustrate a broader point to a general audience.* This common use of reporting involves the wish to be convincing, what Jerome Bruner calls "the myriad expectations that we early, even mindlessly, pick up from the culture in which we are immersed" (2002, p. 65). In order to meet these expectations, therapists will often leave out important information, such as when the report was written, if the reporting was shared with the patient, the length of the treatment, and from what stage of the treatment the material comes (Prochaska, 2000). Without these orienting markers, the reporting is impossible to verify (Spence, 1982).
2. *When therapists describe cases to a general audience.* The same is true when therapists present entire treatments or important aspects of them. The presentation will be based on an observer narrative, and the therapist's own role in the treatment may appear more or less as

the invisible hand directing a recorded performance. The unfortunate effect is that the therapist is portrayed more as a puppet master than as a participant.

3. *When therapists use case illustrations in journals and other professional communications.* This audience will expect certain scholarly standards to be met and is prone to look for inconsistencies. However, when a therapist illustrates the presentation with case material, scholarly standards often are ignored and the therapist reveals very little about his or her role in the treatment. Too often, cases are presented as straight narratives that give no indication over what time period the information was collected and what role the therapist had in how the material surfaced. The patient is presented as completely transparent and responding to a set procedure. At a minimum, the reader/listener should be able to place the vignette in the context of what the patient sought help for and at what point in the treatment the various pieces of information surfaced.

4. *When case reports are used in training therapists.* In this situation, reports are used to assess the knowledge and skill levels of a therapist-in-training. To serve an evaluative purpose, these reports are often long and detailed and their emphasis is on describing entire treatments: how a working alliance was established and how the therapist-in-training dealt with ruptures and impasses, critical turning points, and the process of termination. The format for these reports is a chronological narrative; thus, what Jerome Bruner calls a *paradigmatic narrative* (2002). At the same time, attention is usually given to approaches that are particular to the orientation of the institution responsible for the training. The tendency is, therefore, for therapists-in-training to leave out descriptions of their own role in the treatment and replace them with conceptual constructs. In order to avoid having to show where a clinical mistake may have occurred, a broad, nonspecific language is often favoured. Reports that are part of training thus easily become observer stories or demonstrations of theoretical brilliance, but they fail to be realistic reports.

In all these instances, what is presented needs a common format for describing relational events as experienced by two persons in a serious dialogue. This means that an interactive rather than a theory-based language will more accurately reflect what occurs in a treatment.

Six dimensions of a treatment

Certain dimensions of the therapeutic endeavour are particularly important when reporting. These dimensions become clear when one takes into account working memory and how it functions in treatments. Some of these dimensions may be difficult for the therapist to describe, since much of this information is only fully available during a session, but attention to the six dimensions will no doubt increase the accuracy of the reporting. All these dimensions play an important role in any treatment, and a good case report will touch on most of them. The six dimensions are:

1. *The patient's explicit communication.* The majority of what the therapist remembers from a session is directly related to the patient's explicit verbal communication; thus, most reporting is based on this source. From what we know about working memory, this process is by no means simple and direct but involves several stages (see also Chapter Seven). What the patient communicates must initially be held in the therapist's auditory buffer and the temporary storage of phonological information must be converted into sentences. Although the patient's speech, as most speech, often receives preferential treatment, the processing of it demands more or less undivided attention. The therapist may not hear the entire sentence or may be distracted by other sources of information. Thus, certain information will inevitably be lost. In the second stage, involving the therapist's executive functions, the patient's explicit speech must be processed into narratives that can be stored in long-term memory. To reach this level, the information must be experienced as fully understood by the therapist; it also has to be experienced as relevant to the treatment. This very selective process allows only a small fraction of the information to be retained. What the therapist is able to remember must first activate his or her previously learned scripts and propositions in order for it to be consolidated into long-term memory.

2. *The therapist's explicit communication.* The same stages of memorising must be at work in the patient's processing of the therapist's speech. But since the reporting of what happens in treatments comes from therapists, the information we have is mostly indirect, often related to therapists' experience of being in their own therapy. We have to assume that what patients perceive and remember from their

treatments is similar to the dimension of the therapist's memory of their patients' communication. We must also assume, however, that the patient's attention will differ from that of the therapist. After all, a sense of vulnerability is inevitable when seeking the help of someone unknown who is also cast in the role of expert and given certain authority. Attention to the therapist's speech, although the speech enters into the patient's phonological loop instantly, will vary greatly and will often be quite selective.

3. *The nonverbal communication of the patient.* This is information that requires a different kind of attention and it is often less obvious how the therapist is able to process it. On the one hand, therapists' training places strong emphasis on developing skills in reading the patient's movements, gestures, pauses, and tones of voice (Binder, 2004; Cozolino, 2004). According to Allan Baddeley's (1993) research on working memory, some of this information enters what he calls the *visuospatial sketchpad*, in which the data are quickly dispersed, since spacial input is stored separately from object input. Thus, certain nonverbal information, such as movements and gestures, will be processed as spacial input. On the other hand, the processing of eye contact and tone of voice may rely on other neural mechanisms (ibid.). What appears critical, however, is the therapists' ability to change their focus of attention to include these cues from their patients, while also continuing to process verbal and explicit communication. During a patient's emotionally charged communications, for instance, nonverbal processing appears especially significant.

4. *The nonverbal communication of the therapist.* What the patient perceives nonverbally about his or her therapist's privately held thoughts and feelings is often ignored when discussing what happens in treatments. Louis Cozolino, in his book *The Making of the Therapist* (2006), stresses that through experiencing the therapist's listening, patients are confirmed of the former's presence and attention. This interest is communicated "through [the therapist's] body language and facial expressions" (p. 21). He also stresses the role of eye contact and how it aids in the bonding process. Most experienced therapists appreciate the importance of what they communicate nonverbally to their patients.

5. *The internal processes of the patient.* Here, we are dealing with the patient's internal processes that take place during the therapeutic

dialogue. There is a fine distinction between what is communicated nonverbally and what is strictly based on internal processes; certain thoughts and scripts may not manifest in gestures or obvious facial expressions and thus must be regarded as internal. The patient may remember something but decide to keep this information to himself/herself, or may go into a period of silence and not reveal what he or she is thinking. This is what we may call *privileged information* and the therapist may perceive the effect of this very private activity but could never assume to know its exact nature without the patient's willingness to reveal it.

6. *The internal processes of the therapist.* The internal processes of the therapist have a similar privileged status. Those therapists who primarily take a listening role with their patients no doubt keep more information private. The use of the couch in traditional psychoanalysis, for instance, makes the internal processes of the therapist more difficult for the patient to perceive. Donald Spence (1982) calls this a *privileged competence*, always incomplete and only available for a limited time. As such, it may never enter the therapist's long-term conscious memory, although it often influences the type of interpretations he or she will offer. Still, some of the therapist's internal processes will be perceived by the patient, though what is perceived is bound to be fragmented and often misinterpreted. For instance, the therapist may have no memory of having made certain statements about which the patient is expressing anger. Only if both parties are open to a careful examination of what took place may they discover the source of the issue: what the patient thought the therapist was thinking.

The explicit communication

Unless the therapist is distracted or sleepy, he or she can generally remember most of the topics in a session if cued immediately after its ending (dimension 1). When reviewing a session, the therapist may only remember portions of the patient's dialogue, but we must assume that, from moment to moment, he or she is consciously processing the patient's explicit communication.

The same is probably true about what patients remember immediately after a session (dimension 2). However, exit interviews conducted as part of psychotherapy research often reveal critical gaps in what patients were able to process about their therapist's communication (Matarazzo & Patterson, 1986). Moreover, the accuracy of what is recalled will vary from patient to patient, often related to the degree of psychological stress the patient experienced.

Similar gaps occur in what therapists remember about what they said in a session and, with the passage of time—even hours after a session—these gaps increase considerably (dimension 2). The same is probably true about patients' memory of what they themselves said in a session, although we have very little information about this particular facet of dimension 1.

Psychodynamic theories, in particular, rest on the assumption that therapists have developed explicit ways of observing how patients perceive them. According to these claims, psychoanalysts can detect a patient's preconceived perception of the therapist. This may happen via explicit communication, but often these claims are based on implicit, nonverbal signs of how the therapist is perceived. In making assertions about these preconceived perceptions, called *transference*, analytic practitioners assume that such transference is an inevitable phenomenon in treatments. In particular, transference is assumed to parallel the patient's experience of his or her relationship with the therapist and the patient's significant early parental relationships (Luborsky et al., 1993).

There is no clear verification of this central concept, however. When the psychotherapy research team led by Lester Luborsky attempted to develop a reliable operational method for identifying and measuring transference phenomena, they found that they had to rely on outside raters who would categorise the nature of the interactions by listening to previously taped sessions. To Luborsky and his team, what emerged from these independent assessments was a common understanding of transference phenomena based on clinical judgement rather than on patients' self-reports alone. "The heart of the observations", they write, "is found in the pervasiveness of certain ideas throughout the session and particularly across the narratives in the session" (1993, p. 326).

Similarly, in *Textbook of Psychoanalysis*, Anna Ursula Dreher (2005) of Goethe University, Frankfort, Germany perhaps inadvertently confirms the broadly subjective use of the designation *transference*. She argues against operationalising the term since these attempts would not do justice to the complexity of transference phenomena. At the same time, she appears to agree that transferences express themselves narratively as well as behaviourally. When using the term, Dreher concludes, "an analyst subsumes a number of clinical perceptions and experiences", thus there is "no behavior that as such *is* transference" (p. 364; emphasis in original).

This definition deviates considerably from Freud's formulation that the sign of a patient's transference is his or her resistance to treatment and "the stoppage … of a patient's free associations" (1912b, p. 101). However, both definitions assume that the therapist relies on his or her observations of repeated behavioural cues. Thus, what is meant by the term *transference* must be based on how the therapist interprets nonverbal communication (dimension 3) (Wachtel, 2008).

In reality, we know very little about the nature of each patient's perception of his or her therapist. Episodes may, in fact, occur in a session in which the patient transfers feelings belonging to a child's experiences of its parents, such as the well-rehearsed telling of events in the past, stories that are meant to please and impress a parent. Other episodes may also occur in which this is not the case—in which the patient perceives aspects of the therapist's nonverbal communication or the therapist's inner processes without such distorting influences (dimension 4) (Wachtel, 2008).

Psychotherapy researchers have tried to overcome this problem by using audiovisual recordings, and certain further observations have been made after the fact. However, how the patient perceives the therapist's nonverbal communication is difficult to establish by these methods alone; we have no way of knowing how the therapist's inner processes are perceived unless the patient volunteers this information. Even then, we must assume that the reporting is only partial and not very reliable. In fact, dimensions involving the patient's perception of what the therapist communicates (dimensions 2, 4, and 6) are still in need of studies.

Beyond speech and hearing

Most examples of how therapists process nonverbal communication seem to involve both directly verbal and implicitly nonverbal dimensions. As in the case of Arnold (Chapter Seven), certain aspects of the patient's nonverbal communication led the therapist to hypothesise about Arnold's behaviour toward him. Thus, most of this interpretation was based on elements beyond speech and hearing (dimension 3). The therapist also took notice that Arnold called him by his first name and calls this an "overly chummy" behaviour. Thus, there were concurrent elements of explicit verbal communication (dimension 1) (Strupp & Binder, 1984, p. 82). What allowed the therapist to notice both the verbal and nonverbal cues was, nevertheless, a script or proposition that the therapist was able to call upon. The ability to relate to Arnold's behavioural and verbal communications thus depended on what the therapist had learned and maintained in memory from previous experiences.

More of the patient's nonverbal communication can be brought back if the therapist is interviewed immediately after each session (a technique often used by psychotherapy researchers). But, even then, much is already lost. Only what is particularly significant to the therapist's treatment narrative will be remembered long term. And since nonverbal information is processed via different neural pathways than the dialogue itself, *what the patient communicates nonverbally is only perceived when it can be linked to a preexisting narrative*. The explicit memory of what has already occurred in the treatment is one such narrative that the therapist may cue to process new and nonverbal information.

In psychoanalytic theory, the therapist's awareness of a patient's internal processes is often, post Freud, conceptualised as *countertransference* (dimension 5) (Racker, 1957; Winnicott, 1947).[10] This development has caused some conceptual difficulties. According to Otto Kernberg, two distinct schools of thinking have emerged: under the classical approach, countertransference is "the unconscious reaction of the psychoanalyst to the patient's transference" (1965, p. 38). The other school of thought views it as "the total emotional reaction of the psychoanalyst to the patient in the treatment situation" (ibid.).

The first view corresponds to what the founders of psychoanalysis thought. To them, countertransference was a sign that the analyst's personal issues were interfering with the treatment. For instance, Freud states that "no psycho-analyst goes further than his own complexes and internal resistance permit; and we consequently require that he shall begin his activity with a self-analysis and continually carry it deeper while he is making his observations on his patients" (1910d, p. 145). Thus, problems with countertransference, as the number of practising psychoanalysts increased, became the reason for a training analysis to be part of the therapist's education. The more recent understanding is that countertransference reactions are induced by the patient, thus being acted out with the therapist, who should be able to use these reactions and enactments as tools for exploring experiences that, via the therapist, belong to the patient (Mitchell, 1993).

Both these views ignore much of the therapist's own internal world. Feelings induced by the patient can only account for some of these internal processes (Eagle, 2001). Since the therapist is a participant in the treatment, the nature of this participation must play a significant role in the outcome. On the other hand, viewing countertransference as an unwelcome influence by the therapist's unconscious feelings ignores the fact that the therapist, as a participant, is being influenced by the patient's mood, behaviour, and thoughts. The therapist's responses, even those that remain part of his or her internal processes, cannot altogether be shielded from the patient's perceptions.

Possibly for these reasons, empirical investigations of countertransference have encountered several obstacles (Vakoch & Strupp, 2000). One solution, to develop a treatment manual that could be taught to therapists before conducting controlled studies, appears to have had mainly negative results. As Douglas Vakoch and Hans Strupp of Vanderbilt University discovered, "[M]any of the therapists failed to improve significantly in their ability to deal effectively with essential transference and countertransference issues within actual sessions" (2000, p. 270).

An important reason for this and other failures in using manual treatment guides is pointed out by Jeffrey Binder, at the time one of the researchers involved in the Vanderbilt project. He concludes that what he calls *declarative knowledge*—what is being taught in the classroom about principles and procedures of psychotherapy—does not provide guidance "about when and how to implement concepts, principles, rules, and procedures" (2004, p. 256). Instead, according to Binder, the therapist "may be a 'locker-room expert,' who can 'talk a good game' but who is an inept player" (ibid.).

Binder's comments confirm that the therapist's semantically oriented skills play a secondary role in treatments. When taking other memory systems into account, a far more relevant asset is the therapist's ability to communicate about what is encoded and retrieved via episodic memory. What a treatment manual can accomplish will remain static and may have little to do with skills used in therapy (Binder, 2004). In fact, most treatment manuals are an outgrowth of the academic investigations of psychotherapy outcomes and the need to standardise the data generated. As Binder points out, as training tools these manuals stop short of helping therapists to develop their own narratively based way of remembering and communicating with their patients. The episodic and nonverbal material that is only accessible in the context of ongoing sessions—the therapist's privileged information (as well as the patient's)—will remain out of reach of recording devices, postsession interviews, and other tools used by the researchers.

Memory, memory, memory

In the end, what can be reported about a particular psychotherapy treatment will not meet the standards used for conducting research. When considering all six dimensions of a treatment, each therapeutic encounter is unique and cannot be replicated. Procedures in treatment manuals will never be able to reflect this complexity. For this reason, what therapists report about the treatments in which they participate must always be regarded as a reconstruction—and a reconstruction by only one of the participants in a two-person exchange. As such, a case report describes only what the therapist was able to remember, at a certain point in time, about some select particulars in a treatment—much of which will inevitably become less and less available over time.

Even in notes taken shortly after a session and when describing a therapeutic process, therapists must construct some type of narrative in order to remember and organise what they experienced. Narratives are, therefore, the basis for how therapists report on their work and what therapists remember about it. In fact, much of the ongoing communication with a patient relies on how well these narratives can hold memories for what has occurred in previous interactions.

CHAPTER NINE

Listening in a different state of mind

In answering the question "What is psychotherapy?" Jerome Frank—who for over fifty years researched the many facets of its practice—focuses on two things that psychotherapy is *not* (Frank & Frank, 1991).[1] For one, it is not informal help that may be had from a casual acquaintance, family member, or even a stranger. Furthermore, it is not a medical or surgical procedure. In contrast to bodily intervention, psychotherapy relies on symbolic communication. It is guided by a theory that explains the sources of the patient's distress and prescribes methods for alleviating it (Frank, 2006). He explains:

> An important consequence of the primacy of communication as the medium of healing is that the success of all forms of psychotherapy depends more on the personal influence of the therapist than on medical or surgical procedures. Even when the success of psychotherapeutic procedures is believed to depend solely on their objective properties, as some behaviour therapists maintain, the personal influence of the therapist determines whether the patient carries out the prescribed treatment in the first place, as well as having healing effects in itself.

> (ibid., p. 60)

Frank's conclusion is echoed by the outcome research discussed in the previous chapter (Ahn & Wampol, 2001; Teyber & McClure, 2000). The psychotherapy literature is filled with paradigmatic formulations, procedures, and techniques—what the researchers call "claims of special ingredients". Numerous studies refute the importance of these ingredients and point to the role of the therapist and the relationship between patient and therapist (the therapeutic alliance) as the most significant variables (Lambert, Shapiro, & Bergin, 1986; Lambert & Barley, 2003). Yet, the training of psychotherapists—be it in an academic setting or in freestanding institutes, analytically or from a cognitive-behavioural perspective—still focuses on paradigmatic formulations and techniques. Although there are many forms of psychotherapy—such as working with groups, families, and individuals; short term and long term, and so on—none of them have a much-needed emphasis on how the therapist's memory guides the therapeutic process and how, in turn, the patient's integration of memories is a critical element in treatments.

In reviewing the writings of Freud and his "Recommendations for Physicians on the Psycho-Analytic Method of Treatment" (1912e), now over a century old, the significance of the therapist's memory is nevertheless clearly stated. In fact, it can be argued that his discoveries about how the therapist's memory works during sessions are still relevant to most forms of treatment. When Freud issued what he called "the fundamental rule of psychoanalysis", he was not simply discussing instructions for patients (p. 112). In following his dictum about what is required from the patient, many of his followers seemed to have overlooked what he is saying about the use of one's memory as a therapist.

The term Freud used for his memory functions during sessions, *unconscious memory*, is, in today's perspective, misleading.[2] However, a closer scrutiny of his writings reveals an often precise description of what memory researchers today call *episodic memory* (Schacter, 1996; Tulving, 1972). According to this reading, Freud downplays remembering semantically, via the most common form of conscious memory, while placing major emphasis on remembering episodically. He does so by stressing that the problem in treatments is not how to keep in mind "all the innumerable names, dates, detailed memories and pathological products which each patient communicates in the course of months and years of treatment, and of not confusing them with similar material produced by other patients under treatments simultaneously or previously" (1912e, p. 111). Instead of focusing on such details, Freud insists that the therapist's memory for what happened in each treatment depends on context and retrieval cues—what we know today as hallmarks of episodic memory.

> Those elements of the material which already form a connected context will be at the doctor's conscious disposal; the rest, as yet unconnected and in chaotic disorder, seems at first to be submerged, but rises readily into recollection as soon as the patient brings up something new to which it can be related and by which it can be continued.
>
> (ibid., p. 112)

What connects elements in the patient's disclosures with each other is also the narrative that develops in the therapist's mind and through which he or she remembers each treatment as a process. But for this narrative to develop, Freud recommends that the therapist "should withhold all conscious influences from his capacity to attend", without note taking and without concentrating on anything in particular (ibid.). This is the way that a distinct narrative about each patient develops. After all, he admits, "it must be born in mind that exact reports of analytic case histories are of less value than might be expected" and they "do not succeed in being a substitute for his [the therapist's] actual presence at an analysis" (p. 114).

Thus, the therapist's episodic memory for what has transpired in the treatment, whether it is related to what occurred earlier in the same session or at the beginning of the therapy, is how the therapist contributes to the patient's process. It is what makes the therapist present to the patient, as an engaged witness and trained participant. Freud states this as: "He should simply listen, and not bother about whether he is keeping anything in mind" (1912e, p. 112).

When translated into the mind state that makes episodic retrieval possible, we may today use the term *autonoetic awareness* or *autonoetic knowledge* (these terms are used interchangeably by memory researchers) in which first-person memories of past events can be accessed (Tulving, 1972, 2002). In other words, by entering a reflective state—thus listening with nothing particular in mind and with a distinctly different awareness—the therapist will be cued to his or her personal experiences with the patient.

Listening and conscious memory

Freud's personal method of remembering implies an understanding of memory that emerged many years later. For most psychologists, the fact that memory consists of several systems was for a long time an unreasonable assumption. As discussed in Chapter Two, the difference between semantic and episodic memory remained unknown until certain neural pathways were discovered (Tulving, 2002). Before scientists studied the brains of patients suffering from serious lesions, a unitary understanding of memory prevailed. But with the new findings a principle of what Schacter and colleagues call *class inclusion* developed: impaired performance within a particular memory system appears to have little or no effect on performance outside its class or domain (Schacter, Wagner, & Buckner, 2000).

As two clearly identified, and separate, conscious memory systems, the semantic and episodic memory systems nevertheless share many features. No encoding may in fact be possible in one of the systems without such encoding's also occurring in the other. Thus, the difference between the semantic and episodic memory systems appears to lie in the retrieval process. The research team of Mark Wheeler, Donald Stuss, and Endel Tulving describes the similarities as follows:

> Both episodic and semantic memory can register and hold information about various states of the world, including the internal states of the individual; information stored in both forms of memory is flexibly accessible, being elicitable by a variety of queries, prompts, and cues. Despite this flexibility, not possessed by procedural memory, both episodic and semantic memory obey principles of encoding specificity and transfer appropriate processing.
>
> (1997, p. 333)[3]

In many instances, as the Tulving team concludes, performance on standard tests by healthy subjects is enhanced when both forms of conscious memory can be accessed. Memory of the learning experience—episodic remembering—thus facilitates semantic remembering that functions independently of the specific encoding experience (Eichenbaum, 2002).

However, episodic memory is unique in that it assumes a sense of time travel (Tulving, 2002). Under optimal conditions, contextual cueing will allow the specific experience in which a memory was acquired to be reexperienced, though for semantic memory the time and place of the original experience will no longer be accessible (Eichenbaum, 2002). Another unique characteristic of episodic memory is that it requires a particular state of mind and the ability to withdraw attention from the immediate sensory environment—what Tulving calls *autonoetic knowledge* (Tulving, 1972, 2002). However, when reinforced and consolidated by emotions and heightened attention, the encoded information appears to resist decay. The information will then endure long term, especially when the memory traces can be identified as originating in one's personal experiences (Tulving, 1983; Wheeler, 2000).

The significance of being able to enter this type of awareness is also what Daniel Siegel of the UCLA School of Medicine focuses on in his book *The Mindful Therapist* (2010a). Siegel bases his discussion on neuroscience and applies recent findings from the research to clinical experience. Citing the Tulving group's findings that the orbitofrontal cortex is the key to autonoetic awareness, Siegel finds proof for what he had earlier termed *mindsight*.[4] He later elaborates on these concepts as follows:

> Again, we can imagine the flow of energy as neural firing patterns to one another as it makes maps of past, present, and future in this self-knowing state. We have just identified each of these areas of the middle prefrontal cortex. Orbitofrontal, medial, cingulate and ventrolateral (which includes the anterior insula) zones each play a role in objectivity and self-awareness. (You can see why it is much more efficient to refer to this mindsight-rich area as the middle prefrontal region. It is this area that is activated and strengthened as we become more mindful.)
>
> (p. 111)

Mental time travel is, therefore, what makes mindfulness possible. Siegel proposes that it is "in the chunking of past, present, and future that we come to ride the stream of constructed conceptual awareness" (ibid., p. 231). We may also "imagine a self that is, in fact, a concept of our own construction" (ibid.). The term *insight*, in this respect, stands for the ability to focus attention beyond sensations of the present—into the past as well as into a desired future. Although Siegel does not tie such mindsight to any particular memory function, it is, in fact, closely related to autonoetic awareness.

A similar view, based on the research on autonoetic awareness, is held by Mark Wheeler (2000) of Temple University. The ability to reflect on one's past, present, and future assumes intact prefrontal lobes. One of the benefits of such intact lobes and their sophisticated and reflective awareness is episodic memory: remembering personal experiences in the past. Patients with large prefrontal lesions and children between the ages of one and five years appear to lack this access. Wheeler writes:

> While neither population is amnesic, several different researchers have independently remarked that both groups do not seem to have the same kind of mental life as healthy, adult humans. They do not merely show a dissociation between semantic and episodic memory of the past; they appear to be more generally unable or unwilling to reflect upon their subjective experiences in the past, present, or future.
>
> (p. 606)

An example of the therapist's entering autonoetic awareness is when the patient, Laura, asks, "Did I tell you?", referring to something she discussed in the previous session. Initially, nothing comes to the therapist's mind, but when she says that her daughter still sees a school counsellor, the therapist is cued to a scene about what was discussed in the last session. This discussion may initially have appeared tangential, but it was part of a more central theme of marital problems and raising a child. What Laura was referring to was, in fact, an incident in which her husband passively disagreed with her way of communicating with their daughter and Laura's ability to address his aggression. What on the surface appeared to be a desire to discuss the daughter's counselling now can be placed in the broader context of the patient's marital problems and her struggle to assert herself with her husband.

Listening and relating

Based on what we now know about memory, it appears that Freud, in his technique paper, describes the most critical element in the skill set therapists bring to treatments: the cumulative memory for their direct experiences with a patient and the unique narrative that develops from these experiences. However, he does not explain how to train this skill set. Instead, he focuses on how the therapist must abstain from any emotional disclosures and "must turn his own unconscious like a receptive organ towards the transmitting unconscious of the patient" and "adjust himself to the patient as a telephone receiver is adjusted to the transmitting microphone" (1912e, pp. 115–116).

Such bending of the therapist's attention, Freud incorrectly assumes, will be accomplished by the therapist's working in the manner of a surgeon "who puts aside all his feelings, even his human sympathy, and concentrates his mental forces on a single aim of performing the operation as skilfully as possible" (p. 115). However, such detachment will inevitably compromise the therapist's emotional involvement, a critical component in remembering episodically.

The memory function that most accurately can be called *unconscious* is not related to autonoetic awareness and first-person memory. Rather, it consists of implicit procedural memory and, generally speaking, this type of memory will not function under the control of the therapist's intentions (see Chapter Two). Thus, memory for what occurred specifically in a treatment will be difficult to retain. What is retained will appear mechanical and impersonal, as does most implicit information.

In contrast, when Freud focuses on his own memory and way of listening, he makes no comparison to surgical procedures, as in the quote above. The listening he describes in that personal context is of a different kind, specific to the one-to-one relationships between a therapist and a patient. However, it involves being cued to specific memory material and responding, verbally and nonverbally, with a sense of having understood the deeper meaning of what is being disclosed. This type of cueing is highly dependent on being in a session with the patient and on temporal and special reminders; its retrieval is tied to the context of treatments and the relationship between a specific therapist and a specific patient.

In his striving to teach his listening method to future therapists, Freud thus relied on an educational philosophy that was not equipped to train therapists to listen in a different state of mind. As his reference to surgery makes clear, he assumed that the training of a therapist's autonoetic awareness and episodic memory could be done the same way that training in medicine typically is conducted: as a codification of specific, repeatable procedures. Most of Freud's followers, therefore, appear to have stayed with the common lexical meaning of *listening*, "to make an effort to hear something" (*American Heritage Dictionary*, 1985). Although mechanically correct, this definition certainly misses the point.

In some instances, Freud also appears to have abandoned his own listening method and instead relied on his rule-determined recommendations. As Louis Breger (2000) of the Institute of Contemporary Psychoanalysis points out in reviewing Freud's own cases, only with some patients was he able to approach the treatment with a listening attitude. In other instances, he seems to have held to technique and remained content to interpret what he deemed to be the patient's resistance at the expense of listening autonoetically, with nothing in mind.

The instances in which he appears to have ignored his own rules and instead was an open and receptive participant in the treatments—thus paying more attention to what his memory conveyed—are usually the cases with successful outcomes (ibid.). In cases where he seems to have relied on a theoretical understanding and ignored the relational context, as in the cases of Ida Bauer (Dora), Joseph Wortis, and Clarence Oberndorf, the outcome is more questionable. Breger concludes:

> In none of the cases we have considered was a successful outcome due to what Freud believed was essential to psychoanalytic cure: interpretation of the unconscious, and, most particularly, interpretations derived from his theoretical doctrines. Sometimes, insights into the unconscious were of value in providing meaning for previously bewildering aspects of the patient's experience, though in none of the successful cases was this the major factor.
>
> (2000, p. 372)

To Breger, this can only mean that the psychoanalytic process itself has a curative power "quite separate from interpretation and insight" (ibid.). This process, as therapeutic processes more generally, must in some fundamental ways be tied both to how the therapist accesses episodic memory and to how he or she is able to relate this material in a direct but tactful and empathic way to the patient. When Freud was able to stay personally involved in the latter way, he was a successful therapist. However, when he used his rigid theoretical and technical rules with his own patients, he often failed as a therapist (ibid.).

Differing perspectives

Among those who adhered to Freud's technical recommendations, the descriptions of how to listen therapeutically were, unfortunately, often ignored. And such descriptions did not fare much better with clinicians who found his recommendations too rigid. Those clinicians who followed Jung's example, abandoning many of the specific rules for how to conduct an analysis, continued to produce theoretical formulations, albeit of a different kind (Mitchell & Black, 1995; Schwartz, 1999). Thus, theories continued to multiply, while very little interest was shown in the therapist's memory (Carter & Cambray, 2004; Wilkinson, 2006).

This does not mean that all references to the role of a therapeutic alliance are missing in Jung's formulations. Without specifically mentioning memory, Jung describes the working relationship between himself and a patient as something that developed beyond his medical authority and had little to do with "the outcome of previous psychological experience" (1935a, p. 5). Instead, according to Jung:

> But since all life is to be found only in individual form, and I myself can assert of another individuality only what I find in my own, I am in constant danger either of doing violence to the other person or of succumbing to his influence. If I wish to treat another individual psychologically at all, I must for better or worse give up all pretensions of superior knowledge, all authority and desire to influence. I must perforce adopt a dialectical procedure consisting in a comparison of our mutual findings. But this becomes possible only if I give the other person a chance to play his hand to the full, unhampered by my assumptions.
>
> (ibid.)

This approach, which is consistent with Jung's emphasis on inevitable countertransference, is perhaps one of the earliest descriptions of the therapeutic alliance and an attempt to find what all types of psychotherapy have in common. The "mutual findings" Jung describes are, in today's perspective, also a reference to episodic memory as first-person and subjective, specific to the lived experience between patient and therapist. In insisting on these dialectical formulations, Jung also rejects "all strictly technical methods" as depending on suggestion, procedures, and theorems (1935a, p. 6).

As I have discussed elsewhere, this perspective remained at variance with the prevailing medical model's emphasis on procedures and techniques (Ekstrom, 2002a). And since Jung's descriptions of what he called "a dialectic procedure" had no clear way to be translated into technique, his followers generally focused on a holistic and intrapsychic approach in which "an ongoing dialectic between conscious and unconscious positions" in the patient is explored via dream interpretation (Whitmont & Perera, 1989, p. 7).

How listening has been taught

If Freud's descriptions of how to listen therapeutically were lost on his followers, his emphasis on technique was not. More and more emphasis became placed on knowing the correct formulations and applying proper techniques. This has taken several forms in how training is conducted today; but, overall, very little attention has been given to the particular memory functions therapists rely on and how those functions can be trained.

Three tools dominate in the training of therapists:

- personal therapy, or *training analysis*
- supervision by a senior therapist
- instruction in clinical assessments and psychopathology.

In his book *Freud and His Followers*, Paul Roazen (1975), a historian of psychoanalysis at York University in Toronto, describes the evolution of these three training tools from direct apprenticeship to formal institute training conducted by several senior analysts. In the process, discrepancies between the oral teachings of Freud and his written recommendations developed at an increasing rate.

As noted, training was initially informal; it mainly involved personal analysis and having Freud's approval (Breger, 2000).[5] Becoming a practising psychoanalyst and member of its exclusive organisations was determined through a form of apprenticeship involving first being a patient and undergoing treatment with Freud (ibid.). This meant having a firsthand experience of how Freud worked as a therapist. And, as Roazen writes, Freud was not afraid of being himself: "To some patients Freud seemed the most silent of men, yet compared to later analysts he was almost garrulous. He was capable of being irritatingly quiet and of not saying a word, but he could also chatter; on the whole, he was not as silent as most of today's orthodox Freudians" (1975, p. 121).[6] Inevitably, according to Roazen, two traditions developed:

> During Freud's lifetime his followers were aware of the distinction between "the living personality and the oral teaching of Freud and that of the rigid printed rules" though many tended to adhere to the latter. Since his death this trend has become more pronounced and analysts are much more likely to follow his written recommendations instead of his actual practices.
>
> (p. 119)[7]

As the psychoanalytic movement grew, the need for formal training meant that clinical supervision and classroom instruction were added to the earlier training model. By the late 1920s, the Vienna Psychoanalytic Society had instituted its own training procedures and classes were given by others than Freud (Roazen, 1975). A similar institute was established in Berlin in 1920 and was soon followed by institutes in many other Western countries (Eisold, 2001). Personal analysis, now called *training analysis*, was no longer regarded as therapeutic analysis; further, supervision was still a matter of consulting with senior analysts about problems and therapeutic impasses (Roazen, 1975). Major emphasis was placed on correct theoretical beliefs; in Vienna, new trainees were invited to meetings of the society within a few months so that they could particulate in the presentations and discussions by senior analysts.

Similar systems of training were followed by other schools of analysis. As Thomas Kirsch of the San Francisco Jung Institute describes in his book *The Jungians* (2000), training initially consisted of a personal analysis with Jung; his wife, Emma; or Toni Wolff. This was augmented by didactic exposures. As a professor at the technical university in Zürich (ETH), Jung gave seminars and weekly lectures attended by those interested in training. However, Jung resisted starting his own training institute until after World War II, instead encouraging standards for training in psychotherapy on a nonsectarian basis in Switzerland (ibid.). Formal supervision was minimal, and often the personal analyst would serve in a supervisory capacity as well.

Most of the early psychoanalysts were trained medical doctors, but not all. The U.S. requirement of a medical background was never instituted in Europe or by the many independent training programs that broke with the American Psychoanalytic Association and its various local societies. In fact, academic standards were generally looked upon negatively in most of the early training programs, as was research on outcomes and effectiveness of analytic methods (Roazen, 1975). When treatment became more broadly accepted, however, this was bound to change. Licensing of psychotherapists based on academic degrees became the standard in most countries and, with it, the knowledge base required before receiving training. As training also became available in university settings, the schools of psychotherapy grew to include many nonanalytic orientations.

As Ruth Matarazzo and David Patterson describe in their contribution to *Handbook of Psychotherapy and Behavior Change* (1986), research on what is effective in teaching therapeutic skills took place only when therapists with a background in psychology began exploring ways to standardise instruction and supervision in the 1950s. Unfortunately, most of this research focused on arcane and secondary capabilities such as empathy, warmth, and genuineness.

One thing stands out in how training developed within all of the psychodynamic traditions. The followers of the pioneers—Freud, Jung, Klein, and others—were eager to adhere to the strictest standards and did not want to deviate from what their masters approved. However, a considerable gap was bound to develop between the theoretical formulations of the founders of these "schools" and how they actually worked as therapists. A school's paradigm may refer to the formulations of its founder, but, in reality, the training in and practice of the methods of the founder has undergone considerable development (Safran & Muran, 2000). In the process, Freud's insights into how his memory functioned in treatments have never received the attention they deserve.

Especially in the United States, in attempts to make all psychiatry psychoanalytic, the established training institutes continued to add prerequisite credentials to entering training as well as more rules for those in it (Shorter, 1997). At the same time, these institutes expanded the use of their methods to newer patient populations; there seemed to be no limit to what psychoanalytic therapy could accomplish (Gay, 1998). According to Roazen, "Psychoanalysis grew so fast as a movement that it has sometimes oversold itself as therapy" (1975, p. 171). This observation, made in the 1970s when *Freud and His Followers* was first published, should have raised concerns among psychodynamic therapists—especially since research projects meant to support basic tenets failed to prove superior outcomes (Wallerstein, 2001).

The first training tool: learning from being a patient

The first training tool, *training analysis*, is based on the idea that the psychotherapist's skills are learned from being a patient. This tool, which appears to have been part of psychoanalytic training from its inception, soon came to be the primary educational tool in teaching therapeutic skills. Although originally voluntary when one became a member of Freud's inner circle, these personal analyses were soon mandatory. Jung instituted them in Zürich in 1912; by 1926, training analysis was adopted as official policy of the International Psychoanalytic Association.

The requirement appears to have been a response to some followers' therapeutic failures. (The term *countertransference* was now coined to describe unwelcome and destructive involvement on the analyst's part.) It was never clearly stated, however, how personal analysis would facilitate training and help in developing the skills necessary to be a therapist. As Paul Roazen writes, it was more of an unexamined belief that analysts could be made free of neurotic tendencies and that this would enable them to be more successful with their patients. Thus, the idea of personal analysis was "born of the uncertainties and insecurities of the early analysts" and also promoted the perception of Freud as "a faultless god" (1975, p. 169).

This also meant that therapeutic failures would be attributed to inappropriate or deviant technique and not to lack of listening skills. A rigid and intolerant approach to members who did not adhere to Freud's recommendations resulted in purges and expulsions, often of the most talented analysts. The emphases on proper technique and correct theoretical stances have not been supported by research. Rather, as Jerome and Julia Frank conclude, citing the results of research by the Luborsky team at Penn State University:

> Research provides support for the claim that personal qualities of therapists make appreciable contributions to their success. A repeated finding is that success rates of therapists differ widely and that these differences seem to depend more on the therapist than on the type of treatment.

> (1991, p. 166)

In this perspective, the required personal analysis has turned into a tool to reinforce conformity. Presumably meant to correct the therapist's problems with countertransference, this tool, in the hands of senior analysts, was often used to indoctrinate (Frawley-O'Dea & Sarnat, 2001).[8] It has, therefore, had the devastating effect on training in many independent institutes by creating an educational environment in which the success or failure of the future therapist's personal analysis becomes the entry ticket into professional membership—or rejection from it—after years of education. Especially since the training analysts are often assigned and thus cannot be changed by the candidate, these master analysts become too powerful and often abuse this power.

If the purpose of training analysis is to familiarise the therapist-in-training with a particular method through direct experience, such analysis can never be the same as "real-life" treatments in which the therapists-in-training conduct their own analytic treatments. And if the purpose is to enhance self-knowledge, personal analysis may, in the end, produce mere facility with the terms and theoretical constructs of a certain institution rather than a well-integrated self-narrative in which major life experiences are reflected (Frank & Frank, 1991).

When considering the therapist's memory, a far more important reason for therapists to undergo treatment is to explore their autobiographical memory, either before or during the training. In so doing, the future therapist will have firsthand experience of how to access autonoetic awareness and the more complex forms of retrieval involving the working self.

As we saw in Chapter Three, there are several reasons why we should treat autobiographical memory as different from, or at least as an extension of, episodic memory. Since retrieval of what occurred in the past assumes autonoetic awareness and a reflective state of mind, it appears to demand more time and effort than the retrieval of some basic facts stored semantically (Gazzaniga, 2008). And for the retrieval to result in a sense of expanded self-knowledge, the rememberer would have to confront the fact that the current self may be involved in different life circumstances and have different goals and aspirations than the remembered self (Conway, 2002). In the end, a unified knowledge structure has to emerge, along with a narrative about a uniquely experienced history (Kihlstrom & Klein, 1997).

This, of course, makes the retrieval particularly demanding. The concern is no longer recalling a dinner date, the name of a place visited, or a concert attended. True autobiographical knowledge seems to involve a cohesive sense of self and depends on the capacity to maintain and continually update a narrative about this self (Siegel, 2003). This capacity, as we saw in Chapters Two and Three, is a rather late acquisition in a child's development.

It is difficult, if not impossible, to imagine that a therapist could assist patients in developing a fuller and more complete self-narrative without having had firsthand experience of how this narrative emerges and how it changes basic assumptions about oneself. However, personal analysis may be a more effective educational tool when all ties to the training institutions are removed, when it is guided by the strictest confidentiality, and when therapists-in-training are free to choose their own therapists.

There is, in fact, well-documented support that it is important for therapists to have personal experience that treatments work and of how they work. Patients need to trust that their therapists believe in how they practise; their therapists must project confidence in being able to help. Based on the outcome research, Jeremy Safran and Christopher Muran, two relational psychoanalysts, claim that this trust is the most important part of a therapeutic alliance between patient and therapist in that "it highlights the fact that at a fundamental level the patient's ability to trust, hope, and have faith in the therapist's ability to help always plays a central role in the change process" (2000, p. 13).

Psychotherapy researchers call this *allegiance*, the therapist's belief that the therapy is efficacious. Bruce Wampold of the University of Wisconsin–Madison describes allegiance as follows:

> One of the sacrosanct assumptions of a client is that their therapist believes in the treatment being delivered. Because psychotherapy is an endeavour based on trust, violation of this assumption would appear to undermine the tenets of the profession. For the most part, practising therapists choose the approach to psychotherapy that is compatible with their understanding and conceptualization of psychological distress and health, the process of change, and the nature of the client and his or her issues.
>
> (2001, p. 159)

One way therapists develop a belief in the type of practice they claim is by experiencing treatment themselves. The training analysis is, then, an important form of experience that is difficult to duplicate. However, being in treatment and conducting treatments are not the same. If a training analysis was meant to provide all the necessary experience for candidates to become competent therapists, it is missing one crucial ingredient: the experience of conducting treatments.

The second training tool: learning from another therapist

The second training tool, *practising under supervision*, appears to be an educational tool well suited for learning associated with conducting treatments. Unfortunately, however, there is no clear consensus about how supervision should be given and what purpose it will serve. Jeffrey Binder of the Georgia School of Professional Psychology writes:

> Although authors of recent books on supervision attempt to address the teaching of technical strategies and tactics, they invariably revert to this familiar territory of encouraging the personal growth of trainees and minimizing their narcissistic injuries.
>
> (2004, p. 263)

The "familiar territory" that Binder mentions is the legacy of training within independent institutes. It places the supervisor in an evaluative role; thus, the emphasis is less on facilitating a productive relationship with the patient and more on judging the therapist-in-training's overall psychological maturity. This use of supervision is unlikely to help the therapist-in-training to develop the skill of listening in a different state of mind.

Other approaches to supervision are focused on teaching the application of technique. However, much of the research about what is effective in psychotherapy points to common factors rather than technique. When considering this research, Michael Lambert of Brigham Young University and Benjamin Ogles of Ohio University conclude that "common factors seem to be more important than technique to outcome and therefore should not be neglected in research on manual-based therapy supervision" (1997, p. 434).

More specifically, in studies trying to determine therapeutic competence, the overall conclusion is that experienced therapists, regardless of specific training and approach, show better results than novices. For instance, in referring to a study by Larry Beutler, Marjorie Crago, and Thomas Arizmendi, Jerome and Julia Frank claim that experience is "the crucial determinant of therapeutic competence" (1991, p. 163). To them, "experience correlates with success because psychotherapy is an art that improves with practice" (ibid.).

The timing of supervision. There is also a problem with the timing of the supervision. Commonly, there is a delay of several days, perhaps weeks, during which many of the trainee's memory traces recede. Moreover, when basing the supervision on audio-taped sessions, nonverbal information will be missing. Within hours, most therapists-in-training will also find it difficult to account for their own internal processes while listening to a patient, perhaps the most obvious sign of being able to enter autonoetic awareness.

With the advent of treatment protocols and manuals, many psychotherapy researchers assumed that the problem of accounting for the therapist's internal processes had been solved. But, as Bruce Wampold, himself a seasoned researcher in this field, points out in discussing such manuals, "[T]here is no compelling evidence that [such] adherence is important" (2001, p. 183). Treatment manuals no doubt standardise treatment, thus making it possible to compare different forms of treatment, but since they are based on a group design, treatment manuals may also ignore critical variables, such as the differences in effectiveness among therapists (ibid.). Wampold concludes:

> There has never been a direct comparison between a treatment practised in the clinical trial context (training, manual, supervision, and monitoring) and the same treatment delivered in the clinical practice setting (no extra training, no manual, and no monitoring). In addition, there has never been a comparison of an empirically supported treatment ... and eclectically practised treatment in the clinical practice context.
>
> (p. 213)

The therapist's competent use of episodic memory is clearly another of those variables deserving study. Research focusing on how therapists remember and how they use their memory would help establish how skills related to autonoetic awareness could be trained. This would also tell us what type of supervision would be most beneficial in developing these skills. As long as supervision is geared toward teaching theory, critical memory functions will not be given the attention they deserve.

Teaching the skill of listening and remembering. Episodic memory creates certain limitations of its own. Its highly context-dependent quality may elude the beginning therapist, who instead relies on memory functions used in everyday situations, from indexing and reminder cues to experiences outside the therapy. For lack of experience, beginning therapists will often be reminded of theoretical formulations specific to their training institutes. Thus, when unable to access autonoetic states, beginning therapists are likely to retrieve only semantically encoded propositions and beliefs. Their responses, as a result, will be abstract rather than empathic, no longer traceable to the original encoding events. Significant previous events and experiences in the therapy will be ignored and cannot be revisited with the patient. (See further discussion and case vignette in Chapter Two.)

Another dilemma for beginning therapists is unfamiliarity with processing nonverbal communication. Responding to information expressed via movements, gestures, pauses, and tones of voice requires a somewhat different attention than that used in everyday conversation. Tuning into patients' nonverbal communication is a skill that therapists develop through experience. It is also a skill that requires being able to listen in a particular way, of being both focused and relaxed.

Instead, beginning therapists will be tempted to respond to the overt communication of a patient. They may feel obliged to give advice about dealing with specific everyday problems or to argue against the merit of certain beliefs the patient expresses. In so doing, they will remain in a *noetic* state of mind, in which only semantically encoded material will enter their consciousness (Tulving, 2002).

In light of these difficulties, supervision should be conducted with as little delay as possible and should focus on the therapists-in-training's ability to review their explicit responses—as well their as internal reactions—to the patient being discussed. Supervisory sessions could then be taped and reviewed in order to consolidate the memory of the therapists-in-training. Follow-up interviews could also be done, using two trainees interviewing each other. Such immediate follow-up would guarantee optimal access to therapists' memories and would place less of a financial burden on trainees.

The third training tool: learning to assess and plan

The third element in the training of therapists is *instruction in assessments and psychopathology*. From the inception of psychoanalysis, instruction in how to conduct initial interviews with a patient was required (Freud, 1913c)—today called *assessments and treatment planning*. And since this involves establishing what brought the patient to seek treatment and if psychotherapy is feasible, the available model was one inherited from medicine; in particular, psychiatry (Schwartz, 1999; Shorter, 1997).

For Freud and the pioneers of analytic approaches, the relationship to establishment psychiatry was ambivalent. For instance, in his recommendations for how to begin treatment, Freud was critical of the ease by which his contemporary psychiatrists arrived at a differential diagnosis. The psychiatrist, according to Freud, "is not attempting to do anything that will be of use" and "merely runs the risk of making a theoretical mistake, and his diagnosis is of no more than academic interest" (1913c, p. 124).

> Where the psycho-analyst is concerned, however, if the case is unfavourable he has committed a practical error; he has been responsible for wasted expenditure and has discredited his method of treatment. He cannot fulfil his promise of cure if the patient is suffering, not from hysteria or obsessional neurosis, but from paraphrenia [an older term for schizophrenia] and he therefore has particularly strong motives for avoiding mistakes in diagnosis.
>
> (p. 344)

Among followers of Freud, these reservations soon gave way to making credentials in psychopathology mandatory (Schwartz, 1999). As psychiatrists joined the movement, Freud's methods offered ways of treating psychological problems for which few options had been available (Roazen, 1975). This meant that psychiatry became a respected specialty in medicine and psychiatrists could move from the asylum and into the more desirable private practice (Shorter, 1997). Even among general practitioners of medicine, psychotherapy became a popular way to practise. Thus, as future therapists, doctors entered training already familiar with the procedures of making a diagnosis and following prescribed methods (Roazen 1975; Shorter, 1997). In spite of Freud's misgivings, a medical model for psychotherapy was the inevitable result (Schwartz, 1999).

These developments have not erased all questions about the applicability of medical thinking to what occurs in psychotherapy. As we saw in the last chapter, some psychotherapy researchers point out that mental disorders as classified in diagnostic manuals are still not fully explained; thus, a correct diagnosis may not guarantee a positive outcome (Wampold, 2001). And since the therapist's competence seems to depend on his or her ability to maintain a therapeutic alliance with the patient, diagnosis and prescribed procedures may not be the most critical elements.

Attempting to solve this problem, Jeffrey Binder speaks of two parallel diagnostic tasks. "One task", he writes, "involves developing an estimate about the degree and quality of psychological disturbance from which the patient suffers" (2004, p. 58). He continues:

> This estimate can be codified with various nosological systems. For example, DSM-IV-TR (American Psychiatric Association, 2000) is a descriptive system of categorizing psychopathology in which clusters of behavioral and/or reportable signs and symptoms are distinguished. Another example, from psychoanalytic theory, is Kernberg's (1984) nosological system of categorizing psychopathology in terms of inferred levels of personality organization.
>
> (ibid.)

According to Binder, these systems, with clear roots in medicine, are nevertheless too broad or abstract to guide therapeutic interventions for any given patient. A second diagnostic task has to be added: determining "the problem situation that is unique for this particular patient at this point in his or her life" (p. 59). As discussed in Chapter Two, this task, based far less on identifying psychopathology, involves the development of a therapeutic relationship, an alliance, with the patient. Thus, the therapist's professional experience, communications skills, and ability to relate to what is unique about the patient are here of equal importance (Frank, 2006).

In the first task—formulating a diagnosis of the patient's psychological disturbance—the therapist will rely on a comparison between certain descriptive criteria of specific psychopathologies and his or her observations. The therapist may also use the common psychiatric checklist of mental functions—the mental status exam. And when the assessment uses an established nomenclature, typically from the *Diagnostic and Statistical Manual of Mental Disorders* (the DSM), a certain internal coherence can be expected (Eells, 2007).

By themselves, however, these procedures will not predict the outcome of a therapy nor its process (Binder, 2004; Wampold, 2001). After all, diagnostic formulations do not necessarily reflect a mutual understanding between patient and therapist (Wallerstein, 2001). In their initial meetings with a patient, therapists must consult many different kinds of memorised narratives, all learned before the therapist develops a distinct narrative about the particular patient being interviewed. What the therapist has learned from formal instruction about carrying out a mental status exam and arriving at a diagnosis will be part of this mix of propositional knowledge (see discussion in Chapter Four).

Although this knowledge will provide the therapist with a range of hypothetical explanations and interpretations, to the patient they may still feel impersonal and abstract. As Safran and Muran (2000) point out, the notion of a therapeutic alliance is based on a two-person psychology; it means that a bond is established between patient and therapist. For this to occur, the therapist must have a distinct sense of the patient's personality, critical experiences, and potential healing and must be able to communicate this sense to the patient. Often, this coincides with the therapist's forming the beginning of a narrative particular to the treatment. To the patient, this means that certain tasks and goals have been negotiated with the therapist and a sense of collaboration has emerged.

For the members of the Boston Change Process Study Group (BCPSG) (2002), the therapeutic alliance occurs at what they term *the moment of meeting*. This is an emergent and intersubjective state in which the environment in the therapeutic relationship changes. The alliance is a signal that "something more" has been established, in which two minds acting together create something new and "what comes into being did not exist before and could not be fully predicted by either partner" (BCPSG, 2005, p. 700).

The two-person nature of this alliance has led to the questioning of how applicable the medical model is to psychotherapy (Wampold, 2001). As we saw above and in Chapter Eight, such model is based on claims that special ingredients, such as proper diagnosis and proven procedures, are responsible for good outcomes. In numerous studies, no support has been found for such claims (Frank & Frank, 1991). Rather, the quality of the therapeutic alliance appears to be a far more significant variable (Ahn & Wampold, 2001; Lambert & Barley, 2002). However, the process itself of establishing as thorough a picture as possible of the patient's problems—one ingredient in the medical model—is often part of building this therapeutic alliance. The patient's trust that the therapist has taken time and care to fully understand his or her difficulties will then be a contributing factor to therapist and patient's being able to work together.

The initial assessment is also when the therapist's narrative for the entire treatment begins to develop. In establishing critical aspects of the patient's history, for instance, the therapist forms the first and tentative story about his or her personality, its developmental hurdles, and its resilience.

Learning from and with the patient

The three predominant and longstanding elements in training therapists—personal therapy, supervision, and instruction in assessments and treatment planning—will not, in themselves, produce competent therapists. When we consider how memory functions, learning from and with each patient (and having the tools to do so) is most critical.

Freud (1912e), as we noted earlier, stressed the therapist's "capacity to attend" without note taking and without concentrating on anything in particular; he emphasised what we have called "listening in a different state of mind", the hallmark of autonoetic awareness (p. 112). With this awareness comes a distinct and unfolding narrative about each patient. Whether it is related to what occurred earlier in the same session or at the beginning of the therapy, this narrative is the therapist's main contribution. It is what makes the therapist present to the patient, as an engaged witness and trained participant.

Most training of psychotherapists today uses learning procedures that stress technique in the context of a particular theoretical or paradigmatic system. The three training elements are geared mostly toward proficiency in assessment, planning, and established interventions based on one of these paradigmatic systems. When considering what today is known about explicit or conscious memory, however, *the ability to retain firsthand memory about each patient must be the main goal in training therapists and must be the criterion for the feasibility of training.* Only by developing competence in listening in a different state of mind will therapists be equipped to fully facilitate healing and change.

Other memory systems, such as semantic and procedural memory—even what some researchers call *emotional memory*—no doubt are necessary as well (Siegel, 2003); and knowledge gained from instruction about assessment, planning, and diagnosis will supply material for these memory functions. Nevertheless, for the task of carrying out psychotherapy treatments, these skills must be regarded as auxiliary to the ability to remember episodically.

From the neurocognitive point of view, learning from the patient involves having access to autonoetic awareness and being able to consolidate what is being memorised episodically. The capacity for this type of awareness has several components, one of which is being able to understand the narrative aspects of remembering. Another component has to do with developing a certain kind of working memory and attention to nonverbal communication.

All of this, however, has to occur within a therapeutic relationship. A trusting relationship with the patient must be maintained, a therapeutic alliance in which the specific relationship is acknowledged. As we have seen, extensive studies of psychotherapy outcomes have pointed this out. Much of the recent development in psychodynamic thinking has also been concerned with this aspect of treatments, but often without examining the theoretical basis for treatments. Without taking into account what we know about memory from neuroscientific research, psychodynamic formulations remain incomplete.

NOTES

Preface

1. See Kraly (2009) for a recent review of neurochemical explanations for several mental disorders that also confirms psychotherapy as a valid treatment. He concludes that "the idea that pharmacotherapy alters brain chemistry to relieve symptoms but that psychotherapy does not alter neurochemistry when relieving symptoms is no longer a fair assumption" (p. 21).
2. This painstaking work will often be quoted in this book. In particular, see Siegel (1999), Cozolino (2002), Wilkinson (2010), and Knox (2011).

Chapter One

1. The phrase "cells that fire together wire together" is a credit to Daniel Hebb, the Canadian psychologist who proposed a synaptic theory of learning and memory in 1949.
2. Synaptic plasticity, also called *long-term potentiation*, was first reported by Timothy Bliss and Terje Lomo. According to Charles Gilbert (1999) in *The MIT Encyclopedia of the Cognitive Sciences*: "At the synaptic level, the hippocampus has become the prime model for changes in synaptic weight, through the phenomenon of long-term potentiation originally

described by Bliss and Lomo (1973), and long-term depression. While it has been presumed that these forms of synaptic plasticity account for the storage of complex information in the hippocampus, the linkage has not yet been established. It is clear, however, that cells in the hippocampus are capable of rapidly changing their place fields as the external environment is altered, and this alteration is associated with changes in effective connectivity between hippocampal neurons" (p. 600). LeDoux (2002) terms the plasticity of the brain *self-assembly* and lists seven principles for it. He writes: "Synaptic connections are adjusted by environmentally driven neural activity in specific neural systems. When these changes occur during early life, they are said to involve developmental plasticity; when they occur later, they are considered as learning" (p. 307).

3. In his detailed exploration of the history and sources of psychoanalysis, the historian Henri Ellenberger prefers the term *dynamic psychiatry*, but also refers to the term *depth psychology*, as claiming to "furnish the key to the exploration of the unconscious mind, with wider application to the understanding of literature, art, religion, and culture" (1970, p. 490). According to Jung (1948c), the term *depth psychology* was coined by the Swiss psychiatrist Eugene Bleuler in the early days of psychoanalysis.

4. In discussing the justification for therapists' making historical judgements solely on the patient's account, Schafer (1983) concludes that "the infant or young child of the remote and reconstructed past is rather more of a hypothetical being lacking in individual 'feel' than a concrete presence, and this is so no matter how vividly and empathically the therapist may imagine this past" (p. 206). This is in sharp contrast to the traditional view that childhood experiences are replicated in the transference to the therapist.

5. Ogden refers to Freud's (1913c) comment that "while I am listening to the patient, I, too, give myself over to the current of my unconscious thoughts" (p. 134). Ogden argues that such reverie has always been an unspoken reason for the psychoanalytic use of the couch (Freud, 1913c, in fact refers to the couch as "a remnant of the hypnotic method out of which psycho-analysis evolved" (p. 133)).

Chapter Two

1. A class action lawsuit by four psychologists ended the training monopoly of the American Psychoanalytic Association in 1989. Other, nonaffiliated institutes did not participate in this exclusion and it was not instituted in Europe (Gifford, 2005).

2. In most instances, Tulving uses the terms *autonoetic knowledge* and *autonoetic awareness* interchangeably. In Tulving and Lepage (2000), the second term appears to have won over. For some reason, however, there is also an argument that awareness is not to be equated with consciousness. As I argue in Chapter Three, certainty about what constitutes consciousness is still missing. The term *awareness* is much less exclusive and will be used throughout this book. See also Wheeler (2000), Tulving's colleague and collaborator, who is more consistent in his use of terms.

3. See Schacter, Wagner, and Buckner (2000) for a description of methodology and research criteria.

4. See, for instance, Fisher and Greenberg (1977) and their review of the scientific evidence for Freud's theories of homosexuality and male/female differences in sex-role development.

5. The assumption here is that empathy, as Kohut describes it, is a conscious and well-defined attitude. The other possibility is that it relates to states of reverie, a less conscious attitude. See discussion in Chapter One and references to Ogden (1997).

6. Research by Nichola Calyton and Anthony Dickinson (1998), and Emma Wood, Paul Dudchenko, and Howard Eichenbaum (1999) indicates that animals such as birds and rats also may have some form of episodic memory.

7. It may be worth noticing that this twenty-four-hour period is the same time period that therapists have come to regard as the *day residue*.

8. Conway (2004) also points out that patients with damage to posterior regions of the brain may lose the ability to generate visual images of the past "because episodic content of autobiographical memories is predominately encoded in visual images" (p. 563).

9. This fact has to do with the short duration of what is kept in working memory. As discussed in the next chapter, the limited capacity of this short-term storage may only be overcome when the material is consolidated into narratives and autobiographical knowledge structures. See also Baddeley (2002).

Chapter Three

1. Nelson, de Haan, and Thomas (2006) attribute major developments of the prefrontal cortex to the advancement of memory skills observed in children six to twelve years old. They find that "the changes that have been observed in memory across the preschool and elementary school years ... are likely not due to further maturation of medial temporal lobe structures, but rather, maturation of frontal lobe structures and

importantly, increased connectivity between the medial temporal lobe and the prefrontal cortex" (p. 88).

2. Refers to studies by Nigro and Neisser (1983), and Robinson and Swanson (1993).

3. The patient's transference to the therapist may in some instances trigger memories of the past.

4. Gazzaniga bases his conclusions primarily on English research by Martin Conway and colleagues. See Conway, Pleydell-Pearce, and Whitecross (2001), and Conway, Pleydell-Pearce, Whitecross, and Sharpe (2002–2003).

5. In his autobiographical account, Jung (1961b) claims that his insight into the significance of the self-concept was the result of his meditative practice of painting mandalas. In this respect, his formulations have things in common with Daniel Siegel's (2010a) formulations about mindfulness as being related to autonoetic awareness and autobiographical memory.

6. Posthumous and only recently published, *The Redbook* (2009) documents Jung's personal dreams, paintings, and thoughts during a period of personal crisis and experimentation.

7. In making this definition of the self, Conway (2004) refers to a social psychology study by Albert Hastorf and Hadley Cantril (1952) and their concept of *selected group perception*. In their experiment, they asked students from two universities, Dartmouth and Princeton, about a football game they watched when the two universities played each other. Their perceptions of the game were highly coloured by which school they attended.

8. The quote is from Siegel (1999, p. 229).

Chapter Four

1. For a discussion of how and when children develop the ability to speak about themselves, see also Chapter Three.

2. For a summary definition, see Usher (1993). For Freud's description of the phenomenon, see Freud (1912b).

3. Raymond Mar describes some possible rules applicable to narratives in general. Referring to Barthes (1982) and Dixon (1996), he writes: "These rules of causation demand that events occur in a constrained, logically coherent order. Episodes and actions that allow for other events must take temporal precedence given the conflation of logical (if x then y), causal (because x then y), and temporal priority (first x then y)" (2004, p. 5).

4. In cases of early childhood abuse and neglect, studies by Teicher (2002); Teicher, Anderson, Polcari, Anderson, Navalta, and Kim (2003); and De

Bellis (2005) have shown impaired growth of the corpus callosum that connects the two brain hemispheres.

5. See Hesse (1999). The Adult Attachment Interview (AAI) is a semistructured, hour-long protocol consisting of eighteen questions and is transcribed verbatim upon completion.

6. We may assume that the process is similar to when a child first develops an autobiographical narrative, in that the self-narrative from a treatment has to be co-created; see Nelson (1993); Nelson and Fivush (2000). For a psychoanalytic perspective, see Roy Schafer (1980, 1983, 1992).

7. To Lester Luborksy (1976), the therapeutic alliance, in fact, has two phases: a Type I alliance involving the patient's initial belief in the therapist as a source of help in a warm and supportive relationship, and a Type II alliance in which faith and investment in the therapy process itself are the main features.

8. See discussion in previous chapter and in Chapter Nine. A definition is also provided in Chapter One.

9. Baumeister and Newman (1995) call the indexing function "links between action, resolution and other narrative elements" (p. 99) and give the example of attending a play-off game in which participants apply previous knowledge to the storyline when describing the event. If the purpose of describing what happens in the game is to show how a particular team won the game, the indexing may simply be "goals"; but if the narrator wants to tell why, "belief" would be the index. The narrative could also have "fulfilment" as its index. It would then attempt to describe the ecstatic feelings that resulted when the narrator's team was victorious.

10. For a discussion of the difference between scripts and stories, see Mandler (1984). (A discussion of scripts, stories, and scenes is also found in Chapter Five of this book.)

Chapter Five

1. Domhoff cites several studies indicating only slight differences between home and laboratory dream reports (1996, pp. 42–43).

2. This research has been described by Gazzaniga on pages 132–133 in his book *The Mind's Past* (1998) and LeDoux in The Emotional Brain (1998) on pages 13–14.

3. Studies with rats by a French team of researchers initially seemed to confirm the hypothesis that REM sleep is critical to establishing long-term memories. Later discoveries contradict these results in humans. Peretz Lavie, senior Israeli sleep researcher, concludes that the only function of REM sleep discovered so far is "as a gate to wakefulness while asleep" (Lavie, 1996, p. 150).

4. Siegel discusses the possibility that consolidation of memory occurs in REM sleep, but he prefaces this as a hypothesis in need of validation. He writes: "In this proposal, interhemispheric integration is essential for memory consolidation. Dreaming, REM sleep, and cortical consolidation become the integrating processes that mediate autobiographical narrative. Blockage of these integrating processes may be seen as the core of unresolved trauma and may be revealed as one form of incoherence in autobiographical narratives" (1999, p. 332).

5. The particular regions making functional contributions to dreaming, according to Solms (1997) are: (1) medial occipital-temporal and inferior parietal regions, (2) basal forebrain pathways, and (3) frontal and temporal-limbic structures.

6. Hobson claims that this phenomenon is particular to dreams but is absent in daytime fantasy (1999, p. 121).

7. See also Bulkeley (2005), in which he discusses Solms's research and its findings.

8. In later formulations by Solms and Turnbull (2002), these findings are reiterated as follows: "[T]he parts of the forebrain involved in the constructions of dreams are the entire *limbic system* (including all the 'limbic' components of the frontal and temporal lobes but excluding the 'higher cognitive' components) as well as most of the *visual system* (excluding visual 'projections' cortex). This implies, among other things, that the brain mechanisms of dreams are the same as those for the basic emotions discussed in Chapter Four" (p. 201).

9. See Bulkeley (2001) for contributions by various authors on the rich tradition of interpreting dreams.

10. An argument can, in fact, be made that Jung's theory of archetypes is an attempt to clarify the role of such symbols in what he calls "collective" narratives. Using amplification to decipher a text with an unknown symbol system is a method Jung incorporated via his interest in philology. In working with dreams, he found this method, together with the patient's associations, most useful.

11. Jung uses several terms to describe this aspect of dreams. His earlier references use the term *prospective and finally oriented* (1913a, p. 238) as interchangeable with *teleological* (1913b, p. 201), in both instances in reference to the work of Alphonse Maeder (1906–1907). In other writings, Jung uses the term *prognostic* (1934a, p. 143). In his later writings, the term again is *prospective* or *final point of view*, and he also broadens the categorization to include dreams that are compensatory and reactive (1948a, p. 246).

12. In several instances in the *Collected Works*, Jung refers to his predictions based on dreams. It is not clear, however, if he relied on his own method of amplification of the dream images to arrive at these predictions.

13. In this book, subtitled *How the Brain Goes Out of Its Mind*, Hobson suggests that dreaming has many similarities with mental disorders, such as psychosis, delirium, panic disorder, and dementia.

14. Domhoff's data are based on a study by Snyder (1970).

15. As I discuss in Chapter 6, research by Stephen LaBerge at Stanford University shows a promising new method of recording information from people while they are dreaming (LaBerge, 1998).

16. Spence (1982) discusses this dilemma as a problem of translating pictures into words and is critical of the use of free association, the standard method in Freudian analysis, when attempting to have patients fully relate their dreams. To Spence, free association makes the dream report more fragmentary, and therapists should not be content to work only with dreams as they are reported. However, Spence never discusses the narrative nature of the dream experience itself and his examples only deal with still pictures: classical French paintings. There is no mention of moving imagery.

17. The terms used for the goal of therapy may differ, from Daniel Siegel's emphasis on *integration* to Heinz Kohut's *self-cohesion*. In both instances, however, the access to self and personal history in the form of a self-narrative appears central.

18. See LaBerge (1998) who defines *lucid dreaming* as being able to "act deliberately on reflection or in accordance with plans decided upon before sleep while simultaneously experiencing a dream world that seems vividly real" (p. 500). Hobson (2001), although describing his own experimentation with lucid dreaming, questions whether it takes place "in real time in REM sleep" and instead is "false dreaming because a significant part of our brain is awake" (p. 160).

Chapter Six

1. Gibbs concludes that the image schema of *force* is central to many of the metaphors based on violent body actions and that such schemas play an important role in metaphorical language use (2005, pp. 123–124).

2. This mapping of a target domain to a source in another cognitive domain can be expressed in a formula as: *source → target*; and, in the example above, as *forces → animals*. Commonly, this formula is also expressed as ANIMALS ARE FORCES, a convention used in this book. The small caps are used to indicate that the wording does not occur in language as such but is the conceptual expression of a metaphor.

3. See Kövecses (2010, p. 4) for a description of the convention established by cognitive linguists for presenting conceptual metaphors. Kövecses includes *animals* among common source domains based on how, via observation, animals relate to body and movement.

4. Both *forces* and *animals* could serve as sources for this metaphor in that they both refer to neural maps representing what the body does, such as perceiving and moving.

5. In his major work on dreams, Freud clearly states that he views the dream as a way to access a pathological idea "back to the elements in the patient's mental life from which it originated" (1900a, p. 100). Free association is therefore a method, the patient is told, of "noticing and reporting whatever comes into his head and being misled, for instance, into suppressing an idea because it strikes him as unimportant or irrelevant or because it seems to him meaningless" (p. 101).

6. Jung appears more interested than Freud in what dreams are trying to communicate. In order to figure this out, he suggests, we must assume that dream symbols are the expression of something that could never be fully translated. He bases this argument on data from cultural anthropology.

7. A "neural network" refers to a system of switches interconnected in such a way that they imitate a particular brain structure or a region of the brain. When using computational models for a single neuron, they are referred to as *nodes*.

8. The occipito-temporo-parietal junction, according to Solms, "performs the highest levels of processing of perceptual information" (2005, p. 7). Thus, in dreams, "the 'scene' shifts from the motor end of the apparatus to the perceptual end" (p. 8).

9. The formula for Lakoff's argument for how the dream is related to its interpretation is: D-M->I, *given K*. Letter D refers to the overt or manifest content of the dream; M to a collection of conceptual metaphors; K to knowledge of the dreamer; and DI to interpretation in terms of the dreamer's life (2001, p. 272). The formula therefore explains how DI, the interpreted meaning of the dream, is related to the dream.

10. Solms also refers to Luria (1973) in his description of this site as essential for "the conversion of concrete perception into abstract thinking, which always proceeds in the form of internal schemes, and for the memorizing of organized experience or, in other words, not only for the perception of information but also for its storage" (p. 74).

11. Correa-Beningfield, Kristiansen, Navarro-Ferrando, and Vandeloise classify *support* as a *complex primitive* in that it "can be understood globally (holistically) in terms of the general needs, i.e., the well-being and even survival, of the child" (2005, p. 350). However, *support* and *containment*, according to their research, are not based solely on perceptual information but include image schemas of *forces* and *control*, as well.

12. Based on neural modelling by Narayanan (1997) and his PhD dissertation, Department of Computer Science, University of California–Berkeley.

Chapter Seven

1. The therapist's memory is not mentioned in most psychotherapy research. For instance, Norcross (2002) deals with several elements of the therapy relationship, such as positive regard, congruence, feedback, repairing alliance ruptures, self-disclosure, the management of counter-transference, and relational interpretation. In none of these is the therapist's memory mentioned.
2. See also Deane (2005) and his discussion of metaphors in which kinetic space plays an important role.
3. The authors explain this limitation on the bases of "constraints of time, motivation, financial resources, and human intellectual capacity" (Strupp & Binder, 1984, p. 80).
4. See also Gay (1998, p. 264).
5. Several proverbs are based on similar scripts, like "save us from our friends" that warns against aggression when someone is overly friendly (see Titelman, 1996).

Chapter Eight

1. See also Fancher (1973) and Sulloway (1979). Freud's "Project" was first published in 1977 as part of *Letters to Wilhelm Fliess: Drafts and Notes, 1887–1902*.
2. See discussion in Conway (2002), in which he suggests that episodic memory in particular "may not extend beyond a sleep period during which, perhaps, some consolidation into AM [autobiographical memory] occurs, as well as some forgetting" (p. 63).
3. All literature quoted in earlier chapters and dealing with psychotherapy and neuroscience give these types of short case vignettes. See, for instance, Cozolino (2006), Knox (2011), and Wilkinson (2010).
4. Frank Margison and Anthony Bateman argue that "good psychotherapy needs to combine art in the sense of intuition and judicious use of subjectivity with robust science" (2006, p. 39). They also find that research knowledge, here referring to psychotherapy research, is "notoriously difficult to embed in regular practice".
5. See also volume 3 of *Sigmund Freud: Collected Papers* (1959) and other publications, such as *Three Case Histories*, edited by Philip Rieff (Simon & Shuster, 1996).

6. Roazen (1975) includes Freud, Adler, Jung, Jones, Ferenczi, and Rank among the pioneers. His historical account is based on numerous interviews and visits to hospitals and professional associations between 1965 and 1969.

7. In his detailed exploration of the history and sources of psychoanalysis, the historian Henri Ellenberger prefers the term *dynamic psychiatry*, but also refers to the term *depth psychology*, as claiming to "furnish the key to the exploration of the unconscious mind, with wider application to the understanding of literature, art, religion, and culture" (1970, p. 490). According to Jung (1948c), the term *depth psychology* was coined by the Swiss psychiatrist Eugene Bleuler in the early days of psychoanalysis.

8. "Everybody has won, and all must have prizes" is the Dodo bird's conclusion at the end of the race in Lewis Carroll's *Alice in Wonderland*. Saul Rosenzweig (1936) of Harvard University is credited with having applied this quote to what now is generally understood about the effectiveness of all psychotherapy approaches. (See also Wallerstein, 2001.)

9. Ahn and Wampold's recommendation is that "the focus of counseling research should be on the process of counseling and on the common factors that have historically interested humanistic and dynamic researchers and clinicians" (2001, p. 255).

10. See discussion of more contemporary uses and perspectives in Safran and Muran (2000).

Chapter Nine

1. Jerome D. Frank of the Johns Hopkins University Medical School began researching psychotherapy in the late 1940s; his book *Persuasion and Healing* from 1961 went through many revisions and updates before it was last published with his daughter, Julia Frank, in 1991.

2. By calling his memory in sessions *unconscious*, Freud may have wished to distinguish what he remembered in sessions from other types of recall. However, what he describes is access to episodic memory and the fact that such assess is highly context dependent.

3. The *encoding specificity principle*, first formulated by Tulving and Watkins in 1975, establishes that a retrieval cue can only be effective when the information in the cue was incorporated in the trace of the target event. In other words, the particular way a person thinks about an event determines the likelihood of later recalling it: the cue must reinstate or match the original encoding.

4. The term *mindsight* was first introduced by Siegel in his book *The Developing Mind* (1999), in reference to studies by Fonagy and Target

at University College–London (Fonagy & Target, 1996). This research focused on how children develop a concept of mind and how they can *"detect that another person has a mind with a focus of attention, an intention, and an emotional state"* (Siegel, 1999, p. 200; italics in original).

5. Of the initial members of the Wednesday Society that preceded the Vienna Psychoanalytic Society, some had brief therapies with Freud, others simply became members of this informal group; thus, no personal analysis appears to have been required for those who joined Freud in those early days of the movement (Roazen, 1975).

6. Based on Roazen's interviews with Mark Brunswick and with Philip Sarasin.

7. Quote from Walter Schmideberg (1947), in: To further Freudian psychoanalysis, *The American Imago,* 4(3): 4.

8. Many independent training institutes today insist on the personal analyst's being bound by confidentiality, and thus prohibited from evaluating a candidate's progress in the training.

REFERENCES

Ahn, H., & Wampold, B. E. (2001). Where oh where are the specific ingredients? A meta-analysis of component studies in counseling and psychotherapy. *Counseling Psychology, 48*: 251–257.

Ainsworth, M. D. S., Blehar, M. C., Waters, E., & Wall, S. (1978). *Patterns of Attachment: A Psychological Study of the Strange Situation*. Hillsdale, NJ: Erlbaum.

Alon, N., & Omer, H. (2004). Demonic and tragic narratives in psychotherapy. In: A. Lieblich, D. P. McAdams, & R. Josselson (Eds.), *Healing Plots: The Narrative Basis of Psychotherapy* (pp. 29–48). Washington, DC: American Psychological Association.

Anderson, S. J., & Conway, M. A. (1997). Representation of autobiographical memories. In: M. A. Conway (Ed.), *Cognitive Models of Memory* (pp. 217–246). Cambridge, MA: MIT Press.

Aristotle (1971). Poetics. In: L. R. Loomis (Ed.), *Aristotle on Man in the Universe*. New York: Gramercy.

Aserinsky, E., & Kleitman, N. (1953). Regularly occurring periods of eye motility and concomitant phenomena during sleep. *Science, 118*: 273–274.

Atwood, G. E., & Stolorow, R. D. (1984). *Structures of Subjectivity: Explorations in Psychoanalytic Phenomenology*. Hillsdale, NJ: Analytic Press.

Atwood, G. E., & Stolorow, R. D. (1993). *Faces in a Cloud: Intersubjectivity in Personality Theory*. Northvale, NJ: Jason Aronson.

Baddeley, A. D. (1986). *Working Memory*. Oxford: Oxford University Press.

Baddeley, A. D. (1993). Working memory and conscious awareness. In: A. F. Collins, S. E. Gathercole, M. Conway, & P. E. Morris (Eds.), *Theories of Memory* (pp. 11–28). Hillsdale, NJ: Laurence Erlbaum.

Baddeley, A. D. (2000). Short-term and working memory. In: E. Tulving and F. I. M. Craik (Eds.), *The Oxford Handbook of Memory* (pp. 77–92). Oxford: Oxford University Press.

Baddeley, A. D. (2002). The concept of episodic memory. In: A. Baddeley, M. Conway, & J. Aggleton (Eds.), *Episodic Memory: New Directions in Research* (pp. 1–10). Oxford: Oxford University Press.

Baddeley, A. D., & Hitch, G. J. L. (1974). Working memory. In: G. A. Bower (Ed.), *The Psychology of Learning and Motivation* (*Volume 8*) (pp. 47–89). New York: Academic Press.

Baddeley, A. D., Thornton, A., Chua, S. E., & McKenna, P. (1996). Schizophrenic delusions and the construction of autobiographical memory. In: D. C. Rubin (Ed.), *Remembering Our Past* (pp. 384–428). New York: Cambridge University Press.

Barthes, R. (1982). Introduction to the structural analysis of narratives. In: S. Sontag (Ed.), *A Barthes Reader* (pp. 251–295). New York: Hill and Wang.

Baumeister, R. F., & Newman, L. S. (1995). The primacy of stories, the primacy of roles, and the polarized effects of interpretive motives: some propositions about narratives. In: R. S. Wyer, Jr. (Ed.), *Knowledge and Memory: The Real Story, Advances in Social Cognition* (*Volume 8*) (pp. 97–108). Hillsdale, NJ: Laurence Erlbaum.

BCPSG (Boston Change Process Study Group) (2002). Explicating the implicit: the local level and the microprocess of change in the analytic situation. *International Journal of Psychoanalysis, 83*: 1051–1062.

BCPSG (Boston Change Process Study Group) (2005). The "something more" than interpretation revisited: sloppiness and co-creativity in the psychoanalytic encounter. *Journal of the American Psychoanalytic Association, 53*(3): 693–729.

BCPSG (Boston Change Process Study Group) (2010). *Change in Psychotherapy: A Unifying Paradigm*. New York: W. W. Norton.

Beutler, L. E., Crago, M., & Arizmendi, T. G. (1986). Therapist variables in psychotherapy process and outcome. In: S. L. Garfield & A. E. Bergin (Eds.), *Handbook of Psychotherapy and Behavior Change* (3rd edn., pp. 257–310). New York: John Wiley.

Binder, J. L. (2004). *Key Competencies in Brief Dynamic Psychotherapy: Clinical Practice Beyond the Manual*. New York: Guilford.

Blechner, M. J. (2001). *The Dream Frontier*. Hillsdale, NJ: Analytic Press.

Bornstein, R. F. (2001). The impending death of psychoanalysis. *Psychoanalytic Psychology, 16*: 3–20.

Bowlby, J. (1969). *Attachment and Loss*. (Volume 1, *Attachment*). London: Hogarth.

Braun, A. R., Balkin, T. J., Wesenten, N. J., Carson, R. E., Varga, M., Baldwin, P., Selbie, S., Belenky, G., & Herscovitch, P. (1997). Regional cerebral blood flow throughout the sleep–wake cycle. *Brain, 120*: 1173–1197.

Braun, A. R., Balkin, T. J., Wesenten, N. J., Gwadry, F., Carson, R. E., Varga, M., Baldwin, P., Belenky, G., & Herscovitch, P. (1998). Dissociated pattern of activity in visual cortices and their projections during human rapid eye movement sleep. *Science, 279*: 91–95.

Breger, L. (1974). *From Instinct to Identity: The Development of Personality*. Englewood Cliffs, NJ: Prentice-Hall.

Breger, L. (2000). *Freud: Darkness in the Midst of Vision*. New York: John Wiley.

Bromberg, P. M. (1998). *Standing in the Spaces: Essays on Clinical Process, Trauma, and Dissociation*. Hillsdale, NJ: Analytic Press.

Brown, S. C., & Craik, F. I. M. (2000). Encoding and retrieval of information. In: E. Tulving & F. I. M. Craik (Eds.), *The Oxford Handbook of Memory* (pp. 93–107). New York: Oxford University Press.

Bruner, J. (1990). *Acts of Meaning*. Cambridge, MA: Harvard University Press.

Bruner, J. (1991). The narrative construction of reality. *Critical Inquiry, 18*(3): 1–21.

Bruner, J. (2002). *Making Stories: Law, Literature, Life*. New York: Farrar, Strauss, and Giroux.

Bruner, J., Olver, R. R., & Greenfield, P. M. (1966). *Studies on Cognitive Growth*. New York: Wiley.

Bulkeley, K. (2001). *Dreams: A Reader on the Religious, Cultural, and Psychological Dimensions of Dreaming*. New York: Palgrave.

Bulkeley, K. (2005). *The Wondering Brain: Thinking about Religion with and beyond Cognitive Neuroscience*. New York: Routledge.

Bunge, M. (1999). *Dictionary of Philosophy*. Amherst, NY: Prometheus.

Cambray, J. (2001). Enactments and amplification. *Journal of Analytical Psychology, 46*: 275–303.

Carter, L., & Cambray, J. (2004). Analytic methods revisited. In: J. Cambray & L. Carter (Eds.), *Analytical Psychology: Contemporary Perspectives in Jungian Analysis* (pp. 116–148). New York: Brunner-Routledge.

Christos, G. (2003). *Memory and Dreams: The Creative Human Mind*. New Brunswick, NJ: Rutgers University Press.

Chwalisz, K. (2001). A common factors revolution: let's not "cut off our discipline's nose to spite its face". *Counseling Psychology, 48*: 262–267.

Cienki, A., & Müller, C. (2008). Metaphor, gesture, and thought. In: R. W. Gibbs (Ed.), *The Cambridge Handbook of Metaphor and Thought* (pp. 483–501). New York: Cambridge University Press.

Clayton, N. S., & Dickinson, A. (1998). Episodic-like memory during cache recovery by scrub jays. *Nature, 39*: 272–274.

Conway, M. A. (2001). The phenomenological records and the self-memory system. In: C. Hoerl & T. McCormack (Eds.), *Time and Memory: Issues in Philosophy and Psychology* (pp. 235–255). Oxford: Clarendon.

Conway, M. A. (2002). Sensory perceptual episodic memory and its context: autobiographical memory. In: A. Baddeley, M. Conway, & J. Aggleton (Eds). *Episodic Memory: New Directions in Research* (pp. 53–70). New York: Oxford University Press.

Conway, M. A. (2004). Memory: autobiographical. In: R. L. Gregory (Ed.), *The Oxford Companion to the Mind* (2nd edn., pp. 562–564). New York: Oxford University Press.

Conway, M. A., & Fthenaki, A. (2000). Disruption and loss of autobiographical memory. In: L. S. Cermak (Ed.), *Handbook of Neuropsychology* (2nd edn., pp. 281–312). Amsterdam: Elsevier Science.

Conway, M. A., & Pleydell-Pearce, C. W. (2000). The construction of autobiographical memories in the self-memory system. *Psychological Review, 107*: 261–288.

Conway, M. A., Pleydell-Pearce, C. W., & Whitecross, S. E. (2001). The neuroanatomy of autobiographical memory. *Journal of Memory and Language, 45*: 493–524.

Conway, M. A., Pleydell-Pearce, C. W., Whitecross, S. E., & Sharpe, H. (2002). Brain imaging autobiographical memory. *Psychology of Learning and Motivation, 41*: 229–264.

Conway, M. A., Pleydell-Pearce, C. W., Whitecross, S. E., & Sharpe, H. (2003). Neurophysiological correlates of memory for experienced and imagined events. *Neuropsychologia, 41*: 334–340.

Correa-Beningfield, M., Kristiansen, G., Navarro-Ferrando, I., & Vandeloise, C. (2005). Image-schemas vs. "complex primitives" in cross-cultural spacial cognition. In: B. Hampe (Ed.), *From Perception to Meaning: Image Schemas in Cognitive Linguistics* (pp. 343–366). Berlin: Mouton de Gruyter.

Coulson, S. (2008). Metaphoric comprehension and the brain. In: R. W. Gibbs (Ed.), *The Cambridge Handbook of Metaphor and Thought* (pp. 177–194). Cambridge: Cambridge University Press.

Covington, C. (1995). No story, no analysis? The role of narrative in interpretation. *Journal of Analytical Psychology, 40*: 405–417.

Cozolino, L. (2002). *The Neuroscience of Psychotherapy: Building and Rebuilding the Human Brain*. New York: W. W. Norton.

Cozolino, L. (2004). *The Making of a Therapist: A Practical Guide for the Inner Journey*. New York: W. W. Norton.

Cozolino, L. (2006). *The Neuroscience of Human Relationships: Attachment and the Developing Social Brain*. New York: W. W. Norton.

Culler, J. (1981). *The Pursuit of Signs: Semiotics, Literature, Deconstruction*. Ithaca, NY: Cornell University Press.

Culler, J. (1997). *Literary Theory*. New York: Oxford University Press.

Damasio, A. R. (1999). Introduction. In: The Editors of *Scientific American, The Scientific American Book of the Brain* (pp. ix–xi). Guilford, CT: The Lyons.

Damasio, A. R. (2003). *Looking for Spinoza: Joy, Sorrow, and the Feeling Brain*. New York: Harcourt.

Damasio, A. R. (2010). *Self Comes to Mind: Constructing the Conscious Brain*. New York: Pantheon.

Dautenhahn, K. (2001). The origins of narrative: in search of the transactional format of narratives in humans and other social animals. *International Journal of Cognition and Technology*, 1(1): 97–123.

Deane, P. D. (2005). Multimodal spatial representation: on the semantic unity of *over*. In: B. Hampe (Ed.), *From Perception to Meaning: Image Schemas in Cognitive Linguistics* (pp. 235–282). Berlin: Mouton de Gruyter.

Dewell, R. (2005). Dynamic patterns of containment. In: B. Hampe (Ed.), *From Perception to Meaning: Image Schemas in Cognitive Linguistics* (pp. 369–393). Berlin: Mouton de Gruyter.

Dixon, M. F. N. (1996). *The Polliticke Courtier: Spenser's "The Faerie Queene" as Rhetoric of Justice*. Montreal: McGill-Queene's University Press.

Doctors, S. (2005). An interview with Robert Stolorow. *Self Psychology News*, 1(3): 1–12.

Dodge, E., & Lakoff, G. (2005). Image schemas: from linguistic analysis to neural grounding. In: B. Hampe (Ed.), *From Perception to Meaning: Image Schemas in Cognitive Linguistics* (pp. 57–91). Berlin: Mouton de Gruyter.

Domhoff, G. W. (1996). *Finding Meaning in Dreams: A Quantitative Approach*. New York: Plenum.

Domhoff, G. W. (2005). Refocusing the neurocognitive approach to dreams: a critique of Hobson versus Solms debate. *Dreaming, 1*: 3–20.

Domhoff, G. W. (2010). The case for a cognitive theory of dreams. Available at: http://dreamresearch/net/Library/domhoff_2010.html.

Dreher, A. U. (2005). Conceptual research. In: E. S. Person, A. M. Cooper, and G. O. Gabbard (Eds.), *Textbook of Psychoanalysis* (pp. 361–372). Washington, DC: American Psychiatric Press.

Eagle, M. N. (2001). A critical evaluation of current conceptions of transference and countertransference. *Psychoanalytic Psychology, 17*: 24–37.

Edelson, M. (1992). Telling and enacting stories in psychoanalysis. In: J. W. Barron, M. N. Eagle, & D. L. Wolitzky (Eds.), *Interface of Psychoanalysis and Psychology*. Washington, DC: American Psychological Association.

Eells, T. D. (2007). History and current status of psychotherapy case formulations. In: T. D. Eells (Ed.), *Handbook of Psychotherapy Case Formulation* (2nd edn., pp. 3–32). New York: Guilford.

Eichenbaum, H. (2002). *Cognitive Neuroscience of Memory: An Introduction*. Oxford: Oxford University Press.

Eichenbaum, H., & Bodkin, J. A. (2000). Belief and knowledge as distinct forms of memory. In: D. L. Schacter & E. Scarry (Eds.), *Memory, Brain, and Belief* (pp. 176–207). Cambridge, MA: Harvard University Press.

Eisold, K. (2001). Institutional conflicts in Jungian analysis. *Journal of Analytical Psychology, 46*: 335–353.

Ekstrom, S. R. (2002a). Psychoanalysis and the deviant Jungians: the medical model and our divisive history. *Journal of Jungian Theory and Practice, 3*: 35–52.

Ekstrom, S. R. (2002b). A cacophony of theories: contributions towards a story-based understanding of analytic treatments. *Journal of Analytical Psychology, 47*: 339–358.

Ekstrom, S. R. (2004). The mind beyond our immediate awareness: Freudian, Jungian, and cognitive models of the unconscious. *Journal of Analytical Psychology, 49*(5): 657–682.

Ellenberger, H. F. (1970). *The Discovery of the Unconscious: The History and Evolution of Dynamic Psychiatry*. New York: Basic Books.

Fabricius, J. (1995). The mutual search for the self in the analytic dyad. *International Journal of Psycho-Analysis, 76*: 576–589.

Fancher, R. E. (1973). *Psychoanalytic Psychology: The Development of Freud's Thought*. New York: W. W. Norton.

Federn, P. (1952). *Ego Psychology and Psychoses*. New York: Basic Books.

Fisher, S., & Greenberg, R. P. (1977). *The Scientific Credibility of Freud's Theories and Therapy*. New York: Basic Books.

Fivusch, R. (1997). Gendered narratives: elaboration, structure and emotion in parent-child reminiscing across the preschool years. In: C. P. Thompson, D. J. Hermann, D. Bruce, J. D. Read, D. G. Payne, & M. P. Toglia (Eds.), *Autobiographical Memory: Theoretical and Applied Perspectives* (pp. 79–104). Hillsdale, NJ: Erlbaum.

Flanagan, O. J. (1989). *The Science of the Mind*. Cambridge, MA: MIT Press.

Flew, A. (1979). *A Dictionary of Philosophy* (rev. 2nd edn.). New York: Gramercy.

Fonagy, P. (2005). Psychoanalytic developmental theory. In: E. S. Person, A. M. Cooper, & G. O. Gabbard (Eds.), *Textbook of Psychoanalysis* (pp. 131–145). Washington, DC: American Psychiatric Press.

Fonagy, P., & Target, M. (1996). Theory of mind and the normal development of psychic reality. *International Journal of Psychoanalysis, 77*: 217–233.

Fonagy, P., Steele, H., Moran, G., Steele, M., & Higgit, A. (1991). The capacity for understanding mental states: the reflective self in parent and child and its significance for security of attachment. *Infant Mental Health Journal, 13*: 200–217.

Foulkes, D. (1966). *The Psychology of Sleep.* New York: Charles Scribner's Sons.

Foulkes, D. (1999). *Children's Dreaming and the Development of Consciousness.* Cambridge, MA: Harvard University Press.

Frank, J. D. (2006). What is psychotherapy? In: S. Block (Ed.), *An Introduction to the Psychotherapies* (4th edn., pp. 59–76). Oxford: Oxford University Press.

Frank, J. D., & Frank, J. B. (1991). *Persuasion and Healing: A Comparative Study of Psychotherapy* (3rd edn.). Baltimore, MD: John Hopkins University Press.

Frawley-O'Dea, M. G., & Sarnat, J. E. (2001). *The Supervisory Relationship: A Contemporary Psychodynamic Approach.* New York: The Guilford Press.

Freud, S. (1895). Project for a scientific psychology. *Letters to Wilhelm Fliess: Drafts and Notes, 1887–1902.* New York: Basic Books, 1977.

Freud, S. (1900a). *The Interpretation of Dreams.* In: J. Strachey (Ed.), *The Standard Edition of the Collected Works of Sigmund Freud* (*Volumes IV, V*). London: Hogarth, 1953.

Freud, S. (1905e). Fragments of an analysis of a case of hysteria. In: J. Strachey (Ed.), *The Standard Edition of the Collected Works of Sigmund Freud* (*Volume VII*) (pp. 3–122). London: Hogarth, 1953.

Freud, S. (1909). Letter to Jung, June 30. In: W. McGuire (Ed.), *The Freud/Jung Letters* (149 F, pp. 238–239). Princeton, NJ: Princeton University Press, 1974.

Freud, S. (1909b). Analysis of a phobia in a five-year-old boy. In: J. Strachey (Ed.), *The Standard Edition of the Collected Works of Sigmund Freud* (*Volume X*) (pp. 3–149). London: Hogarth, 1955.

Freud, S. (1909d). Notes upon a case of obsessional neurosis. In: J. Strachey (Ed.), *The Standard Edition of the Collected Works of Sigmund Freud* (*Volume X*) (pp. 153–257). London: Hogarth, 1955.

Freud, S. (1909–1918). *Three Case Histories: The "Wolf Man", the "Rat Man" and the Psychotic Doctor Schreber.* (P. Rieff, Ed.). New York: Simon & Shuster, 1996.

Freud, S. (1911c). Psycho-analytic notes on an autobiographical account of a case of paranoia (dementia paranoids). In: J. Strachey (Ed.), *The Standard Edition of the Collected Works of Sigmund Freud* (*Volume XII*) (pp. 9–82). London: Hogarth, 1958.

Freud, S. (1912b). The dynamics of the transference. In: J. Strachey (Ed.), *The Standard Edition of the Collected Works of Sigmund Freud* (*Volume XII*) (pp. 97–108). London: Hogarth, 1958.

Freud, S. (1912e). Recommendations to physicians practicing psycho-analysis. In: J. Strachey (Ed.), *The Standard Edition of the Collected Works of Sigmund Freud* (*Volume XII*) (pp. 109–120). London: Hogarth, 1958.

Freud, S. (1913c). On beginning the treatment: Further recommendations on the technique of psycho-analysis. In: J. Strachey (Ed.), *The Standard Edition of the Collected Works of Sigmund Freud* (*Volume XII*) (pp. 121–144). London: Hogarth, 1958.

Freud, S. (1916c). A connection between a symbol and a symptom. In: J. Strachey (Ed.), *The Standard Edition of the Collected Works of Sigmund Freud* (*Volume XIV*) (pp. 339–340). London: Hogarth, 1957.

Freud, S. (1916–1917). Introductory lectures on psycho-analysis I and II. In: J. Strachey (Ed.), *The Standard Edition of the Collected Works of Sigmund Freud* (*Volume XV*) (pp. 3–239). London: Hogarth Press, 1961.

Freud, S. (1917d). Metapsychological supplement to the theory of dreams. In: J. Strachey (Ed.), *The Standard Edition of the Collected Works of Sigmund Freud* (*Volume XIV*) (pp. 217–258). London: Hogarth, 1957.

Freud, S. (1918). From the history of an infantile neurosis. In: J. Strachey (Ed.), *The Standard Edition of the Collected Works of Sigmund Freud* (*Volume XVII*). London: Hogarth, 1957.

Freud, S. (1923b). The ego and the id. In: J. Strachey (Ed.), *The Standard Edition of the Collected Works of Sigmund Freud* (*Volume XIX*) (pp. 3–66). London: Hogarth, 1955.

Friedman, W. J. (2001). Memory processes underlying human chronological sense of the past. In: C. Hoerl & T. McCormack (Eds.), *Time and Memory: Issues in Philosophy and Psychology* (pp. 139–167). Oxford: Clarendon Press.

Gardiner, J. M. (2002). Episodic memory and autonoetic consciousness: a first-person approach. In: A. Baddeley, M. Conway, & J. Aggleton (Eds.), *Episodic Memory: New Directions in Research* (pp. 11–30). New York: Oxford University Press.

Gathercole, S. E. (1997). Models of verbal short-term memory. In: M. A. Conway (Ed.), *Cognitive Models of Memory* (pp. 13–45). Cambridge, MA: MIT Press.

Gathercole, S. E., Conway, M. A., Collins, A. F., & Morris, P. E. (1993). The practice of memory. In: A. F. Collins, S. E. Gathercole, M. A. Conway, & P. E. Morris (Eds.), *Theories of Memory* (pp. 1–10). Hove, UK: Lawrence Erlbaum.

Gay, P. (1998). *Freud: A Life for Our Time*. New York, London: W. W. Norton.

Gazzaniga, M. S. (1998). *The Mind's Past*. Berkeley, CA: University of California Press.

Gazzaniga, M. S. (2008). *Human: The Science Behind What Makes Your Brain Unique*. New York: Harper Perennial.

Gibbs, R. W. (2005). The psychological status of image schemas. In: B. Hampe (Ed.), *From Perception to Meaning: Image Schemas in Cognitive Linguistics* (pp. 113–135). Berlin: Mouton de Gruyter.

Gifford, A. (2005). Psychoanalysis in North America from 1895 to the present. In: E. S. Person, A. M. Cooper, & G. O. Gabbard (Eds.), *Textbook of Psychoanalysis* (pp. 387–449). Washington, DC: American Psychiatric Publishing.

Gilbert, C. (1999). Neural plasticity. In: R. A. Wilson & F. C. Keil (Eds.), *The MIT Encyclopedia of the Cognitive Sciences* (pp. 598–601). Cambridge, MA: The MIT Press.

Ginot, E. (2007). Intersubjectivity and neuroscience: understanding enactments and their therapeutic significance within emerging paradigms. *Psychoanalytic Psychology, 24*(2): 317–332.

Godden, D. R., & Baddeley, A. D. (1975). Context-dependent memory in two natural environments: on land and under water. *British Journal of Psychology, 66*: 325–331.

Grady, J. E. (2005). Image schemas and perception: refining a definition. In: B. Hampe (Ed.), *From Perception to Meaning: Image Schemas in Cognitive Linguistics* (pp. 35–55). Berlin: Mouton de Gruyter.

Grünbaum, A. (1993). *Validation in the Clinical Theory of Psychoanalysis*. Madison, CT: International Universities Press.

Hall, C. S. (1953). A cognitive theory of dreams. In: S. G. M. Lee & A. R. Mayes (Eds.), *Dreams and Dreaming: Selected Readings*. Harmondsworth: Penguin Education, 1973.

Harris, A. (2005). Transference, countertransference, and the real relationship. In: E. S. Person, A. M. Cooper, & G. O. Gabbard (Eds.), *Textbook of Psychoanalysis* (pp. 201–216). Washington, DC: American Psychiatric Publishing.

Hartmann, E., Russ, D., Oldfield, M., Falke, R., & Skoff, B. (1980). Dream content: effects of L-DOPA. *Sleep Research, 9*: 153.

Hastorf, A. H., & Cantril, H. (1952). They saw a game: a case study. *Journal of Abnormal Social Psychology, 49*: 129–134.

Hesse, E. (1999). The adult attachment interview: historical and current perspectives. In: J. Cassidy & P. R. Shaver (Eds.), *Handbook of Attachment: Theory, Research, and Clinical Applications* (pp. 395–433). New York: Guilford Press.

Hesse, E., Main, M., Abrams, Y., & Rifkin, A. (2003). Unresolved states regarding loss or abuse can have "secondary generation" effects: disorganization, role inversion, and frightening ideation in the offspring of traumatized, non-maltreated parents. In: D. J. Siegel & M. F. Salomon (Eds.), *Healing Trauma: Attachment, Mind, Body, and Brain* (pp. 57–106). New York and London: W. W. Norton.

Hillman, J. (1983). *Healing Fiction*. Barrytown, NY: Station Hill. [Originally published in J. Wiggins (Ed.), *Religion as Story*. New York: Harper & Row, 1975.]

Hobson, J. A. (1988). *The Dreaming Mind*. New York: Basic Books.

Hobson, J. A. (1994). *The Chemistry of Conscious States*. Boston and New York: Little, Brown.

Hobson, J. A. (1999). *Dreaming as Delirium: How the Brain Goes Out of Its Mind*. Cambridge, MA: MIT Press.

Hobson, J. A., & McCarley, R. (1977). The brain as a dream state generator: an activation-synthesis hypothesis of the dream process. *American Journal of Psychiatry, 134*: 1335–1348.

Hobson, J. A., Pace-Schott, E. F., & Stickgold, R. (2003). Dreaming and the brain: toward a cognitive neuroscience of conscious states. In: E. F. Pace-Schott, M. Solms, M. Blagrove, & S. Harnad (Eds.), *Sleep and Dreaming: Scientific Advances and Reconsiderations* (pp. 1–50). Cambridge: Cambridge University Press.

Hodges, J. R., & Graham, K. S. (2002). Episodic memory: insights from semantic dementia. In: A. Baddeley, M. Conway, & J. Aggleton (Eds.), *Episodic Memory: New Directions in Research* (pp. 132–152). New York: Oxford University Press.

Hoerl, C. (2001). The phenomenology of episodic recall. In: C. Hoerl & T. McCormack (Eds.), *Time and Memory: Issues in Philosophy and Psychology* (pp. 315–335). Oxford: Clarendon Press.

Horvath, A. O., & Bedi, R. P. (2002). The alliance. In: J. C. Norcross (Ed.), *Psychotherapy Relationships That Work* (pp. 37–69). Oxford and New York: Oxford University Press.

Hudson, J. A., & Nelson, K. (1986). Repeated encounters of a similar kind: effects of familiarity on children's autobiographical memory. *Cognitive Development, 1*: 253–271.

Janowsky, J. S., Shimamura, A. P., & Squire, L. R. (1989). Source memory impairments in patients with frontal lobe lesions. *Neuropsychologica, 27*: 1043–1056.

Johnson, M. (1987). *The Mind in the Body: The Bodily Basis of Meaning, Imagination, and Reason*. Chicago: University of Chicago Press.

Johnson, M. (2005). The philosophical significance of image schemas. In: B. Hampe (Ed.), *From Perception to Meaning: Image Schemas in Cognitive Linguistics* (pp. 15–33). Berlin: Mouton de Gruyter.

Johnson, M. (2007). *The Meaning of the Body: Aestetics of Human Understanding.* Chicago: University of Chicago Press.

Johnson, M. (2008). Philosophy's dept to metaphor. In: R. Gibbs (Ed.), *The Cambridge Handbook of Metaphor and Thought* (pp. 39–52). Cambridge: Cambridge University Press.

Josselson, R. (2004). On becoming the narrator of one's own life. In: A. Lieblich, D. P. McAdams, & R. Josselson (Eds.), *Healing Plots: The Narrative Basis of Psychotherapy* (pp. 111–127). Washington, DC: American Psychological Association.

Jung, C. G. (1913a). General aspects of psychoanalysis. In: W. McGuire (Ed. and Trans.), *The Collected Works of C. G. Jung (Volume 4)* (pp. 229–242). Princeton, NJ: Princeton University Press, 1961.

Jung, C. G. (1913b). The theory of psychoanalysis. In: W. McGuire (Ed. and Trans.), *The Collected Works of C. G. Jung (Volume 4)* (pp. 83–203). Princeton, NJ: Princeton University Press, 1961.

Jung, C. G. (1934a). The practical use of dream-analysis. In: W. McGuire (Ed. and Trans.), *The Collected Works of C. G. Jung (Volume 16, 2nd edn.)* (pp. 139–161). Princeton, NJ: Princeton University Press, 1966.

Jung, C. G. (1934b). A review of the complex theory. In: W. McGuire (Ed. and Trans.), *The Collected Works of C. G. Jung (Volume 8)* (2nd ed., pp. 92–104). Princeton, NJ: Princeton University Press, 1969.

Jung, C. G. (1935a). Principles of practical psychotherapy. In: W. McGuire (Ed. and Trans.), *The Collected Works of C. G. Jung (Volume 16)* (2nd edn., pp. 3–20). Princeton, NJ: Princeton University Press, 1966.

Jung, C. G. (1935b). The relations between the ego and the unconscious. In: W. McGuire (Ed. and Trans.), *The Collected Works of C. G. Jung (Volume 7)* (2nd edn., pp. 123–241). Princeton, NJ: Princeton University Press, 1966.

Jung, C. G. (1935c). The Tavistock lectures. In: W. McGuire (Ed. and Trans.), *The Collected Works of C. G. Jung (Volume 18)* (pp. 5–182). Princeton, NJ: Princeton University Press, 1955.

Jung, C. G. (1946). The psychology of the transference. In: W. McGuire (Ed. and Trans.), *The Collected Works of C. G. Jung (Volume 16)* (2nd edn., pp. 163–323). Princeton, NJ: Princeton University Press, 1966.

Jung, C. G. (1948a). General aspects of dream psychology. In: W. McGuire (Ed. and Trans.), *The Collected Works of C. G. Jung (Volume 8)* (2nd edn., pp. 237–280). Princeton, NJ: Princeton University Press, 1969.

Jung, C. G. (1948b). On the nature of dreams. In: W. McGuire (Ed. and Trans.), *The Collected Works of C. G. Jung (Volume 8)* (2nd edn., pp. 281–297). Princeton, NJ: Princeton University Press, 1969.

Jung, C. G. (1948c). Depth psychology. In: W. McGuire (Ed. and Trans.), *The Collected Works of C. G. Jung (Volume 18)* (pp. 477–488). Princeton, NJ: Princeton University Press, 1955.

Jung, C. G. (1961a). Symbols and the interpretation of dreams. In: W. McGuire (Ed. and Trans.), *The Collected Works of C. G. Jung (Volume 18)* (pp. 185–264). Princeton, NJ: Princeton University Press, 1955.

Jung, C. G. (1961b). *Memories, Dreams, Reflections.* (A. Jaffe, Ed., and R. Winston & C. Winston, Trans.). Rev. edn. New York: Vintage Books, 1965.

Jung, C. G. (2009). *The Red Book.* (S. Shamdasani, Ed.). New York: W. W. Norton.

Jus, A. (1973). Studies on dream recall in chronic schizophrenia patients after prefrontal lobotomy. *Biological Psychiatry, 6*: 275.

Kalsched, D. (1996). *The Inner World of Trauma: Archetypal Defenses of the Personal Spirit.* London: Routledge.

Kernberg, O. F. (1965). Notes on countertransference. *Journal of the American Psychoanalytic Association, 13*: 38–56.

Kernberg, O. F. (1980). *Internal World and External Reality.* New York: Jason Aronson.

Kihlstrom, J. F., & Klein, S. B. (1997). Self-knowledge and self-awareness. *Annals of the New York Academy of Sciences, 818*: 5–17.

Kirsch, T. B. (2000). *The Jungians: A Comparative and Historical Perspective.* London: Routledge.

Knox, J. (2003). *Archetype, Attachment, Analysis.* Hove, NY: Brunner-Routledge.

Knox, J. (2011). *Self-Agency in Psychotherapy: Attachment, Autonomy, and Intimacy.* New York: W. W. Norton.

Kohut, H. (1959). Introspection, empathy, and psychoanalysis. In: P. H. Ornstein (Ed.), *The Search for the Self (Volume 1)* (pp. 205–232). New York: International Universities Press, 1978.

Kohut, H. (1973). *The Restoration of the Self.* New York: International Universities Press.

Kohut, H. (1984). *How Does Analysis Cure?* Chicago: The University of Chicago Press.

Kosslyn, S. M. (1994). *Image and Brain: The Resolution of the Imagery Debate.* Cambridge, MA: MIT Press.

Kövecses, Z. (2010). *Metaphor: A Practical Introduction* (2nd edn.). Oxford: Oxford University Press.

Kradin, R. (2006). *The Herald Dream: An Approach to the Initial Dream in Psychotherapy.* London: Karnac.

Kraly, F. S. (2009). *The Unwell Brain: Understanding the Psychobiology of Mental Health.* New York: W. W. Norton.

LaBerge, S. (1998). Dreaming and consciousness. In: S. R. Hameroff, A. W. Kaszniak, & A. C. Scott (Eds.), *Toward a Science of Consciousness II: The Second Tucson Discussions and Debates* (pp. 495–504). Cambridge, MA: MIT Press.

Lakoff, G. (1987). *Women, Fire, and Dangerous Things: What Categories Reveal about the Mind*. Chicago: University of Chicago Press.

Lakoff, G. (2001). How metaphor structures dreams: the theory of conceptual metaphor applied to dream analysis. In: K. Bulkeley (Ed.), *Dreams: A Reader on Religious, Cultural, and Psychological Dimensions of Dreaming* (pp. 265–283). New York: Palgrave.

Lakoff, G. (2008). The neural theory of metaphor. In: R. Gibbs (Ed.), *The Cambridge Handbook of Metaphor and Thought* (pp. 17–38). Cambridge: Cambridge University Press.

Lakoff, G., & Johnson, M. (1980). *Metaphors We Live By*. Chicago: University of Chicago Press.

Lakoff, G., & Johnson, M. (1999). *Philosophy in the Flesh: The Embodied Mind and Its Challenge to Western Thought*. New York: Basic Books.

Lakoff, G., & Turner, M. (1989). *More Than Cool Reason: A Field Guide to Poetic Metaphor*. Chicago: University of Chicago Press.

Lambert, M. J., & Barley, D. E. (2002). Research summary on the therapeutic relationship and psychotherapy outcome. In: J. C. Norcross (Ed.), *Psychotherapy Relationships That Work: Therapist Contributions and Responsiveness to Patients* (pp. 17–32). New York, NY: Oxford University Press.

Lambert, M. J., & Ogles, B. M. (1997). The effectiveness of psychotherapy supervision. In: C. E. Watkins, Jr. (Ed.), *Handbook of Psychotherapy Supervision* (pp. 421–446). New York: Wiley.

Lambert, M. J., Shapiro, D. A., & Bergin, A. E. (1986). The effectiveness of psychotherapy. In: S. L. Garfield & A. E. Bergin (Eds.), *Handbook of Psychotherapy and Behavior Change* (3rd edn., pp. 157–212). New York: John Wiley.

Lavie, P. (1996). *The Enchanted World of Sleep*. New Haven, CT: Yale University Press.

LeDoux, J. (1995). In search of an emotional system in the brain: leaping from fear to emotion and consciousness. In: M. S. Gazzaniga (Ed.), *The Cognitive Neurosciences* (pp. 1049–1061). Cambridge, MA: MIT Press.

LeDoux, J. (1998). *The Emotional Brain*. London: Weidenfeld & Nicolson.

LeDoux, J. (2002). *Synaptic Self: How Our Brain Becomes Who We Are*. New York: Viking.

Lessem, P. A. (2005). *Self Psychology: An Introduction*. New York: Jason Aronson.

Levine, B., Svoboda, E., Hay, J. F., Winocur, G., & Moscovitch, M. (2002). Aging and autobiographical memory: dissociating episodic from semantic retrieval. *Psychology and Aging, 17*(4): 677–689.

Levine, H. B. (1994). The analyst's participation in the analytic process. *International Journal of Psycho-Analysis, 75*: 665–676.

Lieblich, A., McAdams, D. P., & Josselson, R. (2004). Introduction. In: A. Lieblich, D. P. McAdams, & R. Josselson (Eds.), *Healing Plots: The Narrative Basis of Psychotherapy* (pp. 3–9). Washington, DC: American Psychological Association.

Lippman, P. (2000). *Nocturnes: On Listening to Dreams*. Hillsdale, NJ: Analytic Press.

Loftus, E. F. (1981). Memory and its disortions. In: A. G. Kraus (Ed.), *The G. Stanley Hall Lecture Series* (pp. 123–154). Washington, DC: American Psychological Association.

Luborsky, L. (1976). Helping alliances in psychotherapy. In: J. L. Cleghhorn (Ed.), *Successful Psychotherapy* (pp. 92–116). New York: Bruner/Mzel.

Luborsky, L., Barber, J. P., Binder, J., Curtis, J., Dahl, H., Horowitz, L. M., Horowitz, M., Perry, J. C., Schacht, T., Silberschatz, G., & Teller, V. (1993). Transference-related measures: a new class based on psychotherapy sessions. In: N. E. Miller, L. Luborsky, J. P. Barber, & J. P. Docherty (Eds.), *Psychodynamic Treatment Research: A Handbook for Clinical Practice* (pp. 326–360). New York: Basic Books.

Luborsky, L., Crits-Christoph, P., Mintz, J., & Auerbach, A. (1988). *Who Will Benefit from Psychotherapy? Predicting Therapeutic Outcomes*. New York: Basic Books.

Luria, A. (1973). *The Working Brain*. New York and Harmondsworth: Penguin.

Maguire, E. A. (2002). Neuroimaging studies of autobiographical memory. In: A. Baddeley, M. Conway, & J. Aggleton (Eds.), *Episodic Memory: New Directions in Research* (pp. 164–180). New York: Oxford University Press.

Main, M., & Goldwyn, R. (1998). Adult attachment scoring and classification system. Unpublished manuscript, University of California at Berkeley.

Mandler, J. M. (1984). *Stories, Scripts, and Scenes: Aspects of Schema Theory*. Hillsdale, NJ: Laurence Erlbaum.

Mandler, J. M. (2004). *The Foundation of Mind: Origins of Conceptual Thought*. Oxford: Oxford University Press.

Mandler, J. M. (2005). How to build a baby: III; Image schemas and the transition to verbal thought. In: B. Hampe (Ed.), *From Perception to Meaning: Image Schemas in Cognitive Linguistics* (pp. 137–163). Berlin: Mouton de Gruyter.

Mar, R. A. (2004). The neuropsychology of narrative: story comprehension, story production and their interrelation. *Neuropsychologia, 42*(10): 1414–1434.

Marginson, F., & Bateman, A. (2006). Research in psychotherapy. In: S. Block (Ed.), *An Introduction to the Psychotherapies* (4th edn., pp. 37–57). Oxford: Oxford University Press.

Matarazzo, R. G., & Patterson, D. (1986). Research on the teaching and learning of therapeutic skills. In: S. L. Garfield & A. E. Bergin (Eds.), *Handbook of Psychotherapy and Behavior Change* (3rd edn., pp. 821–844). New York: John Wiley.

Mayes, A. R., & Roberts, N. (2002). Theories of episodic memory. In: A. Baddeley, M. Conway, & J. Aggleton (Eds.), *Episodic Memory: New Directions in Research* (pp. 86–109). New York: Oxford University Press.

McAdams, D. P. (1993). *The Stories We Live By: Personal Myths and the Making of the Self.* New York: The Guilford Press.

McAdams, D. P. (2006). *The Redemptive Self: Stories Americans Live By.* Oxford: Oxford University Press.

McClelland, J. (2000). Connectionist models of memory. In: E. Tulving & F. I. M. Craik (Eds.), *The Oxford Handbook of Memory* (pp. 583–608). Oxford: Oxford University Press.

McGaugh, J. L. (1995). Emotional activation, neuromodulatory systems and memory. In: D. L. Schacter (Ed.), *Memory Distortion: How Minds, Brains, and Societies Reconstruct the Past* (pp. 255–273). Cambridge, MA: Harvard University Press.

McKenna, P., & Warrington, E. K. (2000). The neuropsychology of semantic memory. In: L. S. Cermak (Ed.), *Handbook of Neuropsychology* (2nd edn., pp. 355–382). Amsterdam: Elsevier Science B. V.

McWilliams, N. (1999). *Psychoanalytic Case Formulations.* New York: Guilford Press.

Metcalfe, J. (2000). Metamemory: theory and data. In: E. Tulving & F. I. M. Craik (Eds.), *The Oxford Handbook of Memory* (pp. 197–211). Oxford: Oxford University Press.

Messer, S. B., & Wolitzky, D. L. (1997). The traditional psychoanalytic approach to case formulation. In: T. D. Eells (Ed.), *Handbook of Psychotherapy Case Formulation* (pp. 26–57). New York: Guilford Press.

Messer, S. B., & Wolitzky, D. L. (2007). The traditional psychoanalytic approach to case formulation. In: T. D. Eells (Ed.), *Handbook of Psychotherapy Case Formulation* (2nd edn., pp. 67–104). New York: Guilford Press.

Miller, N. E., Luborsky, L., Barber, J. P., & Docherty, J. P. (Eds.) (1993). *Psychodynamic Treatment Research: A Handbook for Clinical Practice.* New York: Basic Books.

Mitchell, S. A. (1988). *Relational Concepts in Psychoanalysis: An Integration.* Cambridge, MA: Harvard University Press.

Mitchell, S. A. (1993). *Hope and Dread in Psychoanalysis.* New York: Basic Books.

Mitchell, S. A., & Black, M. J. (1995). *Freud and Beyond: A History of Modern Psychoanalytic Thought.* New York: Basic Books.

Narayanan, S. (1997). Embodiment in language understanding: sensory-motor representations for metaphoric reasoning about event descriptions. PhD dissertation, Department of Computer Science, University of California–Berkeley.

Nelson, C. A., de Haan, M., & Thomas, K. M. (2006). *Neuroscience of Cognitive Development: The Role of Experiences and the Developing Brain*. Hoboken, NJ: John Wiley.

Nelson, K. (1993). The psychological and social origins of autobiographical memory. *Psychological Science, 4*: 7–14.

Nelson, K. (2000). Memory and belief in development. In: D. L. Schacter & E. Scarry (Eds.), *Memory, Brain, and Belief* (pp. 259–289). Cambridge, MA: Harvard University Press.

Nelson, K., & Fivush, R. (2000). Socialization of memory. In: E. Tulving & F. I. M. Craik (Eds.), *The Oxford Handbook of Memory* (pp. 283–295). Oxford: Oxford University Press.

Nigro, G., & Neisser, U. (1983). Point of view in personal memories. *Cognitive Psychology, 15*: 465–482.

Norcross, J. C. (Ed.) (2002). *Psychotherapy Relationships That Work: Therapist Contributions and Responsiveness to Patients*. Oxford: Oxford University Press.

Nyberg, L., & Cabeza, R. (2000). Brain imagining of memory. In: E. Tulving & F. I. M. Craik (Eds.), *The Oxford Handbook of Memory* (pp. 501–519). Oxford: Oxford University Press.

O'Connor, A. R., Moulin, C. J. A., & Cohen, G. (2008). Memory and consciousness. In: G. Cohen & M. A. Conway (Eds.), *Memory in the Real World* (3rd edn., pp. 327–356). Hove and New York: Psychology Press.

Ogden, T. H. (1997). *Reverie and Interpretation: Sensing Something Human*. Northvale, NJ: Jason Aronson.

Patterson, K., & Hodges, J. R. (2000). Semantic dementia: one window on the structure and organisation of semantic memory. In: L. S. Cermak (Ed.), *Handbook of Neuropsychology* (2nd edn., pp. 313–333). Amsterdam: Elsevier Science B. V.

Person, E. S., Cooper, A. M., & Gabbard, G. O (Eds.) (2005). *Textbook of Psychoanalysis*. Washington, DC: American Psychiatric Publishing.

Pinker, S. (2002). *The Blank Slate: The Modern Denial of Human Nature*. New York: Viking.

Prochaska, J. O. (2000). Change at different stages. In: C. R. Snyder & R. E. Ingram (Eds.), *Handbook of Psychological Change: Psychotherapy Processes and Practices for the 21st Century* (pp. 109–127). New York: John Wiley.

Pulvermüller, F. (2002). *The Neuroscience of Language: On Brain Circuits of Words and Serial Order*. Cambridge: Cambridge University Press.

Racker, H. (1957). The meaning and use of countertransference. *Psychoanalytic Quarterly, 26*: 303–357.

Ramachandran, V. S. (2004). *A Brief Tour of Human Consciousness*. New York: Pi Press.

Ramachandran, V. S. (2011). *The Tell-Tale Brain: A Neuroscientist's Quest for What Makes Us Human*. New York: W. W. Norton.

Revonsuo, A. (2003). The reinterpretation of dreams: an evolutionary hypothesis of the function of dreaming. In: E. F. Pace-Schott, M. Solms, M. Blagrove, & S. Harnad (Eds.), *Sleep and Dreaming: Scientific Advances and Reconsiderations* (pp. 85–109). Cambridge: Cambridge University Press.

Rice, H. & Rubin, D. C. (2011). Remembering from any angle: the flexibility of visual perspective during retrieval. *Consciousness and Cognition, 20*: 568–577.

Roazen, P. (1975). *Freud and His Followers*. New York: Da Capo Press.

Robinson, J. A., & Swanson, K. L. (1993). Field and observer modes of remembering. *Memory, 1*: 169–184.

Rock, A. (2004). *The Mind at Night*. New York: Basic Books.

Rohrer, T. (2005). Image schemata in the brain. In: B. Hampe (Ed.), *From Perception to Meaning: Image Schemas in Cognitive Linguistics* (pp. 165–196). Berlin: Mouton de Gruyter.

Rolls, E. T. (2000). Memory systems in the brain. *Annual Review of Psychology, 5*: 599–630.

Roussy, F., Camirand, C., Folkes, D., DeKoninch, J., Loftis, M., & Kerr, N. H. (1996). Does early night REM dream content reliably reflect presleep state of mind? *Dreaming, 6*: 121–130.

Ryle, A., & Bennett, D. (1997). Case formulations in cognitive analytic therapy. In: T. D. Eells (Ed.), *Handbook of Psychotherapy Case Formulation* (pp. 289–313). New York: Guilford Press.

Safran, J. D., & Muran, J. C. (2000). *Negotiating the Therapeutic Alliance: A Relational Treatment Guide*. New York: The Guilford Press.

Schacter, D. L. (1996). *Search for Memory: The Brain, the Mind, and the Past*. New York: Basic Books.

Schacter, D. L., Wagner, A. D., & Buckner, R. L. (2000). Memory systems of 1999. In: E. Tulving & F. I. M. Craik (Eds.), *The Oxford Handbook of Memory* (pp. 627–643). Oxford: Oxford University Press.

Schafer, R. (1980). Narration in the psychoanalytic dialogue. In: W. J. T. Mitchell (Ed.), *On Narrative* (pp. 25–49). Chicago: University of Chicago Press.

Schafer, R. (1983). *The Analytic Attitude*. New York: Basic Books.

Schafer, R. (1992). *Retelling a Life: Narration and Dialogue in Psychoanalysis*. New York: Basic Books.

Schank, R. C. (1990). *Tell Me a Story: Narrative and Intelligence*. Evanston, IL: Northwestern University Press.

Schank, R. C. (1999). *Dynamic Memory Revisited*. Cambridge: Cambridge University Press.

Schank, R. C., & Abelson, R. (1977). *Scripts, Plans, Goals and Understanding*. Hillsdale, NJ: Lawrence Erlbaum.

Scholes, R. (1980). Language, narrative, and anti-narrative. In: W. J. T. Mitchell (Ed.), *On Narrative* (pp. 200–208). Chicago: University of Chicago Press.

Schore, A. N. (1994). *Affect Regulation and the Origin of the Self: The Neurobiology of Emotional Development*. Hillsdale, NJ: Lawrence Erdbaum.

Schore, A. N. (2003). *Affect Regulation and the Repair of the Self*. New York, London: W. W. Norton.

Schwartz, J. (1999). *Cassandra's Daughter: A History of Psychoanalysis*. New York: Viking.

Shorter, E. (1997). *A History of Psychiatry: From the Era of the Asylum to the Age of Prozac*. New York: John Wiley.

Siegel, D. J. (1999). *The Developing Mind: How Relationships and the Brain Interact and Shape Who We Are*. New York: The Guilford Press.

Siegel, D. J. (2003). An interpersonal neurobiology of psychotherapy: the developing mind and the resolution of trauma. In: M. F. Solomon & D. J. Siegel (Eds.), *Healing Trauma: Attachment, Mind, Body, and Brain* (pp. 1–56). New York, London: W. W. Norton.

Siegel, D. J. (2007). *The Mindful Brain: Reflection and Attunement in the Cultivation of Well-Being*. New York: W. W. Norton.

Siegel, D. J. (2010a). *The Mindful Therapist: A Clinician's Guide to Mindsight and Neural Integration*. New York: W. W. Norton.

Siegel, D. J. (2010b). *Mindsight: The New Science of Personal Transformation*. New York: Bantam Books.

Snyder, F. (1970). The phenomenology of dreaming. In: L. Madow & L. Snow (Eds.), *The Psychodynamic Implications of Physiological Studies on Dreaming* (pp. 124–151). Springfield, IL: Thomas.

Solms, M. (1997). *The Neuropsychology of Dreams*. Mahwah, NJ: Laurence Erlbaum.

Solms, M. (2000). The interpretation of dreams and neuroscience. The British Psychoanalytic Society (pp. 1–10). (www.psychoanalysis.orguk/solms4.htm)

Solms, M. (2003). Dreaming and REM sleep are controlled by different brain mechanisms. In: E. F. Pace-Schott, M. Solms, M. Blagrove, & S. Harnad (Eds.), *Sleep and Dreaming: Scientific Advances and Reconsiderations* (pp. 51–58). Cambridge: Cambridge University Press.

Solms, M., & Turnbull, O. (2002). *The Brain and the Inner World: An Introduction to the Neuroscience of Subjective Experience*. New York: Other Press.

Sorsoli, L. (2004). Echoes of silence: remembering and repeating childhood trauma. In: A. Lieblich, D. P. McAdams, & R. Josselson (Eds.), *Healing Plots: The Narrative Basis of Psychotherapy* (pp. 89–109). Washington, DC: American Psychological Association.

Spence, P. D. (1982). *Narrative Truth and Historical Truth: Meaning and Interpretation in Psychoanalysis*. New York: W. W. Norton.

Spence, P. D. (1993). Traditional case studies and prescriptions for improving them. In: N. E. Miller, L. Luborsky, J. P. Barber, & J. P. Docherty (Eds.), *Psychodynamic Treatment Research: A Handbook for Clinical Practice* (pp. 37–52). New York: Basic Books.

Spiegel, D. (1995). Hypnosis and suggestion. In: D. L. Schacter (Ed.), *Memory Distortion: How Minds, Brains, and Societies Reconstruct the Past* (pp. 129–149). Cambridge, MA: Harvard University Press.

Squire, L. R. (1987). *Memory and Brain*. Oxford: Oxford University Press.

Squire, L. R., & Zola, S. M. (1996). Structure and function of declarative and nondeclarative memory systems. *Proceedings of the National Academy of Science USA, 93*: 13515–13522.

Sternberg, R. J. (1996). *Cognitive Psychology*. Fort Worth, TX: Harcourt Brace College.

Stolorow, R. D., Brandchaft, B., & Atwood, G. E. (1987). *Psychoanalytic Treatment: An Intersubjective Approach*. Hilldale, NJ: The Analytic Press.

Strupp, H. H. (1986). The nonspecific hypothesis of therapeutic effectiveness: a current assessment. *American Journal of Orthopsychiatry, 56*: 513–520.

Strupp, H. H., & Binder, J. L. (1984). *Psychotherapy in a New Key: A Guide to Time-Limited Dynamic Psychotherapy*. New York: Basic Books.

Stuss, D. T., & Benson, D. F. (1986). *The Frontal Lobes*. New York: Raven Press.

Sulloway, F. J. (1979). *Freud, Biologist of the Mind: Beyond the Psychoanalytic Legend*. New York: Basic Books.

Tagini, A. & Raffone, A. (2009). The "I" and the "me" in self-referential awareness: a neurocognitive hypothesis. *Cognitive Processing, 11*: 9–20.

Teicher, M. (2002). Scars that will not heal: the neurobiology of child abuse. *Scientific American, 286*: 68–75.

Teicher, M. H., Anderson, S. L., Polcari, A., Anderson, C. M., Navalta, C. P., & Kim, D. M. (2003). The neurobiological consequences of early stress and childhood maltreatment. *Neuroscience Biobehavioral Review,* 27(1–2): 33–44.

Teyber, E., & McClure, F. (2000). Therapist variables. In: C. R. Snyder & R. E. Ingram (Eds.), *Handbook of Psychological Change: Psychotherapy Processes and Practices for the 21st Century* (pp. 62–87). New York: John Wiley.

The American Heritage Dictionary (2nd college edn.) (1985). Boston, MA: Houghton Mifflin.

Titelman, G. Y. (1996). *Popular Proverbs and Sayings.* New York: Gramercy.

Tulving, E. (1972). Episodic and semantic memory. In: E. Tulving & W. Donaldson (Eds.), *Organization of Memory* (pp. 381–403). New York: Academic Press.

Tulving, E. (1983). *Elements of Episodic Memory.* New York: Oxford University Press.

Tulving, E. (2000). Concepts of memory. In: E. Tulving & F. I. M. Craik (Eds.), *The Oxford Handbook of Memory* (pp. 33–43). Oxford: Oxford University Press.

Tulving, E. (2002). Episodic memory and common sense: how far apart? In: A. Baddeley, M. Conway, & J. Aggleton (Eds.), *Episodic Memory: New Directions in Research* (pp. 269–284). New York: Oxford University Press.

Tulving, E., & Lepage, M. (2000). Where in the brain is the awareness of one's past? In: D. L. Schacter & E. Scarry (Eds.), *Memory, Brain, and Belief* (pp. 208–258). Cambridge, MA: Harvard University Press.

Tulving, E., & Markowitsch, H. J. (1998). Episodic and declarative memory: role of the hippocampus. *Hippocampus, 8*: 198–204.

Tulving, E., & Thomson, D. M. (1973). Encoding specificity and retrieval processes in episodic memory. *Psychological Review, 80*: 352–373.

Tulving, E., & Watkins, M. (1975). Structure of memory traces. *Psychological Review, 82*: 261–275.

Ulanov, A. B. (1982). Transference/countertransference: a Jungian perspective. In: M. Stein (Ed.), *Jungian Analysis.* La Salle, IL: Open Court.

Usher, S. F. (1993). *Introduction to Psychodynamic Psychotherapy Techniques.* Madison, CT: International Universities Press.

Vakoch, D. A., & Strupp, H. H. (2000). Psychodynamic approaches to psychotherapy: philosophical and theoretical foundations of effective practice. In: C. R. Snyder & R. E. Ingram (Eds.), *Handbook of Psychological Change: Psychotherapy Processes and Practices for the 21st Century* (pp. 200–216). New York: John Wiley.

Van der Kolk, B. A. (1996). Trauma and memory. In: B. A. van der Kolk, A. C. McFarlane, & L. Weisaeth (Eds.), *Traumatic Stress: The Effects of Overwhelming Experience on Mind, Body, and Society* (pp. 279–302). New York: Guilford Press.

Van der Kolk, B. A. (2003). Posttraumatic stress and the nature of trauma. In: M. F. Solomon & D. J. Siegel (Eds.), *Healing Trauma: Attachment, Mind, Body, and Brain* (pp. 168–195). New York: W. W. Norton .

Wachtel, P. L. (1992). On theory, practice, and the nature of integration. In: R. B. Miller (Ed.), *The Restoration of Dialogue: Readings in the Philosophy of Clinical Psychology* (pp. 418–433). Washington, DC: American Psychological Association.

Wachtel, P. L. (2008). *Relational Theory and Practice of Psychotherapy*. New York: Guilford Press.

Wagner A. D., Maril, A., & Schacter, D. L. (2000). Interactions between forms of memory: when priming hinders new learning. *Journal of Cognitive Neuroscience, 12*: 52–60.

Wagner, A. D., Schacter, D. L., Rotte, M., Koutstaal, W., Maril, A., Dale, A. M., Rosen, B. R., & Buckner, R. L. (1998). Verbal memory encoding: brain activity predicts subsequent remembering and forgetting. *Science, 281*: 1188–1191.

Wallerstein, R. S. (2001). The generations of psychotherapy research: an overview. *Psychoanalytic Psychology, 18*: 243–267.

Wampold, B. E. (2001). *The Great Psychotherapy Debate: Models, Methods, and Findings*. New York: Routledge.

Wampold, B. E., Ahn, H., & Coleman, H. L. K. (2001). Medical model as metaphor: old habits die hard. *Counseling Psychology, 48*: 268–273.

Watrous, A. J., Tandon, N., Connor, C. R., Pieters, T., & Ekstrom, A. D. (2013). Frequency-specific network connectivity increases underlie accurate spatiotemporal memory retrieval. *Nature Neuroscience, 16*: 349–356 (DOI;10,1038/nn.3315).

Westbury, C., & Dennett, D. C. (2000). Mining the past to construct the future: memory and belief as forms of knowledge. In: D. L. Schacter & E. Scarry (Eds.), *Memory, Brain, and Belief* (pp. 11–32). Cambridge, MA: Harvard University Press.

Wheeler, M. A. (2000). Episodic memory and autonoetic awareness. In: E. Tulving & F. I. M. Craik (Eds.), *The Oxford Handbook of Memory* (pp. 597–608). Oxford: Oxford University Press.

Wheeler, M. A., Stuss, D. T., & Tulving, E. (1997). Toward a theory of episodic memory: the frontal lobes and autonoetic consciousness. *Psychological Bulletin, 121*: 331–354.

Whitmont, E. C., & Perera, S. B. (1989). *Dreams: A Portal to the Source*. New York: Routledge.

Wilkinson, M. (2004). The mind–brain relationship: the emergent self. *Journal of Analytical Psychology, 49*: 83–101.

Wilkinson, M. (2006). *Coming to Mind: The Mind–Brain Relationship*. London: Routledge.

Wilkinson, M. (2010). *Changing Minds in Therapy: Emotion, Attachment, Trauma and Neurobiology*. New York: W. W. Norton.

Williams, H., Conway, M., & Cohen, G. (2008). Autobiographical memory. In: G. Cohen & M. A. Conway (Eds.), *Memory in the Real World* (3rd edn., pp. 21–90). Hove and New York: Psychology Press.

Winer, R. (1994). *Close Encounters: A Relational View of the Therapeutic Process*. Northvale, NJ: Jason Aronson.

Winnicott, D. W. (1947). Hate in the countertransference. In: M. M. R. Kahn (Ed.), *Through Paediatrics to Psycho-Analysis* (pp. 194–203). New York: Basic Books, 1975.

Winnicott, D. W. (1960). Counter-transference. In: M. M. R. Kahn (Ed.), *The Maturational Processes and the Facilitating Environment* (pp. 158–165). New York: International Universities Press, 1965.

Winnicott, D. W. (1962). The aims of psycho-analytic treatment. In: M. M. R. Kahn (Ed.), *The Maturational Processes and the Facilitating Environment* (pp. 166–170). New York: International Universities Press, 1965.

Wolf, E. S. (1988). *Treating the Self: Elements of Clinical Self Psychology*. New York: Guilford Press.

Wood, E. R., Dudchenko, P. A., & Eichenbaum, H. (1999). The global record of memory in hippocampal neuronal activity. *Nature, 397*: 613–616.

Woody, S. R., Detweiler-Bedell, J., Teachman, B. A., & O'Hearn, T. (2003). *Treatment Planning in Psychotherapy: Taking the Guesswork out of Clinical Care*. New York: Guilford Press.

INDEX

References/Bibliography

EXERCISE

Active Older Adults . . . and the YMCA, can be purchased by writing to the YMCA Program Store, Box 5077, Champaign, IL 61820—price $16.00.

Adult Physical Fitness—A Program For Men and Women. President's Council on Physical Fitness. Superintendent of Documents, U.S. Government Printing Office, Washington, DC, 1963.

Anderson, B. *Stretching*. Bolinas, CA: Shelter Publications, Inc., 1980.

Arthritis Foundation YMCA Aquatic Program, Instructor's Manual. YMCA of the USA and the Arthritis Foundation. Human Kinetics Publishers, Inc., 1985. Copies of the booklet can be purchased from the YMCA Program Store, Box 5077, Champaign, IL 61820.

Basic Exercises For People Over Sixty. National Association For Human Development. 1750 Pennsylvania Ave., N.W., Washington, DC 20006.

Bender, Ruth. *Be Young And Flexible After 30-40-50-60*. Avon, CT: Ruben Publishing Company.

Chrisman, Dorothy C. *Body Recall: A Program of Physical Fitness For the Adult*. Burea, KY: Berea College Press, 1980.

Christenson, A. and Rankin, D. *Easy Does It: Yoga For Older People*. New York: Harper & Row, 1975.

Clarke, Ed. H. Harrison. *Physical Fitness Research Digest: Exercise And Aging*. President's Council on Physical Fitness and Sports, Washington, D.C. 20201.

Cooper, K. *Running Without Fear: How To Reduce the Risk of Heart Attack and Sudden Death During Aerobic Exercise*. New York: M. Evans & Co., 1985.

Cooper, K. *The New Aerobics*. Philadelphia: M. Evans & Co., 1970.

Ebel, H., Sol, N., Bailey, D., and Schechter, S. (Eds.). *Presidential Sports Award Fitness Manual*. Havertown, PA: FitCom Corporation, 1983.

Fletcher, Colin. *The New Complete Walker: The Joys and Techniques Of Hiking and Back Packing*. New York: Knopf, 1974.

Fluegelman, Andrew. *The New Games Book*. Garden City, NY: A Headlands Press, 1976.

Frankel, Lawrence Jr. and Richard, Betty Byrd. *Be Alive As Long As You Live: The Older Person's Complete Guide To Exercise For Joyful Living*. New York: Lippincott and Crowell, 1980.

Golding, Lawrence A., Meyers, Clayton, R., and Sinning, Wayne E. (Eds.), *The Y's Way To Physical Fitness*, National Board of YMCA, Chicago, 1982.

Gray, John. *Racewalking For Fun And Fitness*. New Jersey: Prentice-Hall, Inc., 1985.

Guidelines for Graded Exercise Testing and Exercise Prescription. American College of Sports Medicine. Philadelphia: Lea & Febiger, 1980.

Guild, Warren R. *How to Keep Fit And Enjoy It*. New York: Harper and Row, 1962.

Harris, Raymond and Frankel, Lawrence. *Guide To Fitness After Fifty*. Plenum Press, 1977.

Hurley, Olga. *Safe Therapeutic Exercise For The Frail Elderly: An Introduction*. Center for the Study of Aging, New York, 1987.

Jacobson, Howard. *Racewalk To Fitness*. New York: Simon & Schuster, 1980.

Jamieson, R. H. *Exercises For The Elderly*. New York: Emerson Books, 1982.

Join The Active People Over 60! National Association for Human Development in cooperation with the President's Council on Physical Fitness and Sports, under a grant from the U.S. Administration on Aging.

Keeler, R. *Pep Up Your Life: A Fitness Book For Seniors*. Hartford, CT: The Travelers Insurance Companies, 1980.

Krewer, S. *The Arthritis Exercise Book*. New York: Simon & Schuster, 1981.

Kuntzleman, C.T. *The Complete Book Of Walking*. New York: Simon & Schuster, 1977.

Leslie, David K. and McLure, John W. *Exercises For The Elderly*. Iowa Commission on Aging. University of Iowa Press, 1975.

Lesser, Mercedes. The Effects Of Rhythmic Exercise On The Range Of Motion In Older Adults. *American Correctional Therapy Journal*, July-August 1978, *32*(4).

Melleby, A. *The Y's Way To A Healthy Back*. New Jersey: New Century Publishers, 1982.

Mockenhaupt, Robin E. *Exercise and Fitness, Learning Centers For Health*. National Health Screening Council For Volunteer Organizations, Inc., Washington, D.C., 1980.

Moderate Exercises For People Over Sixty. National Association for Human Development. 1750 Pennsylvania Ave., N.W., Washington, DC

Nideffer, Robert M. *Athletes' Guide To Mental Training*. Champaign, IL: Human Kinetics Publishers, Inc., 1985.

Norton, S. *Yoga For People Over 50*. Old Greenwich, CT: Devin-Adair Company, 1977.

Pearl, Bill and Moran, Gary. *Getting Stronger*. Bolinas, CA: Shelter Publications, 1986.

Pollock, Michael L. "How Much Exercise is Enough?" *The Physician and Sportsmedicine*, June 1978, *6*(6).

Prentice, W.E. and Bucher, C.A. *Fitness for College and Life*. St. Louis, MO: Times Mirror/Mosby College Printing, 1988.

Rodah, Kare. *Be Fit For Life: A Practical Guide To Physical Well-Being*. New York: Harper and Row, 1966.

Scripps Research Foundation. *Cardiovascular Exercise Primer*. La Jolla, CA: Preventive Medicine Center of Scripps Clinic and Research Foundation, n.d.

Sharkey, Brian J. *Physiology of Fitness: Prescribing Exercise For Fitness, Weight Control, and Health*. Champaign, IL: Human Kinetics Publishers, Inc., 1984.

Sheehan, G. *Running and Being: The Total Experience*. New York: Simon & Schuster, 1978.

Smith, E.L. and Gilligan, C. Physical Activity: Prescription For The Older Adult. *The Physician and Sportsmedicine*, 1983, *11*:91-103.

Spackman, Robert R., Jr. *Conditioning For Fitness – Physical Fitness After High School and College.* Carbondale, IL: Hillcrest House, 1979. (For orders: Schwebel Printing, P.O. Box 433, Murfreesboro, TN 62960.)

The Fitness Challenge in the Later Years. President's Council on Physical Fitness and Sports. Superintendent of Documents, U.S. Government Printing Office, Washington, DC, June 1975.

Trent, George D. *The Gentle Art Of Walking.* New York: Knopf, 1974.

U.S. Dept. of Health, Education and Welfare, Administration on Aging. *The Fitness Challenge in Later Years: Exercise Program For Older Americans.* Washington, DC: reprinted June 1975.

Walking For Exercise and Pleasure. The President's Council On Physical Fitness and Sports. Washington, DC 20201.

Warshaw, L.J. *Managing Stress.* Reading, MA: Addison-Wesley, 1979.

Wear, Robert et al. *Fitness, Vitality And You: Serving The Elderly, The Techniques, Part 2.* New England Gerontology Center for Continuing Education, 15 Garrison Ave., University of New Hampshire, Durham, NH, 1977.

Webb, T. *Tamilee Webb's Original Rubber Band Workout.* New York: Workman Publishing Co., 1986.

Wilson, Jr. C.H., Exercise For Arthritis. In Basmajian, J.V. (Ed.) *Therapeutic Exercise* (pp. 514-530). Baltimore, MD: Williams and Wilkins Co., 1978.

Wilson, K. Twinges in the Hinges. Unpublished manuscript. 1983.

YMCA – Physical Fitness Through Water Exercise. Champaign, IL: Human Kinetics, 1982.

Zohman, Lanore R. *Exercise Your Way To Fitness and Heart Health.* American Heart Association and President's Council on Physical Fitness and Sports, 1974.

Zook, S. and Zook K.P. *The Rubber Band Shape-Up Program.* Phoenix, AZ: ORTHO-SPORT Publications, 1987.

Zuti, W.B. *The Official YMCA Fitness Program.* New York: Rawson Associates, 1984.

Zuti, W.B. and Mandelblit, Cole. *YMCA Corporate Health Enhancement Program.* National Board of Young Men's Christian Associations, 1980.

AGING AND RELATED SUBJECTS

Accidents and the Elderly. Age Page. National Institute on Aging, July 1980.

Berlinger, W.G. and Spector, R. Adverse Drug Reactions In The Elderly. *Geriatrics*, 1984, *39*:45-58.

Botwinick, Jack. *Aging and Behavior: A Comprehensive Integration Of Research Findings*. New York: Springer Publishing Company, 1973.

Bristow, R. *Aches and Pains: How The Older Person Can Find Relief Using Heat, Massage, and Exercise*. New York: Random House, 1974.

Britton, Joseph H. and Britton, Joan O. *Personality Changes In Aging: A Longitudinal Study Of Community Residents*. New York: Springer Publishing Company.

Bromley, D.B. *The Psychology Of Human Aging*, Revised Edition. New York: Penguin Books.

Bruh, Wilde, MD. *The Psychology of Human Aging*. New York: W.W. Norton and Company.

Busse, E.W., MD. Eating In Late Life: Physiologic and Psychologic Factors. *New York State Journal of Medicine*, August 1980: 1496-1497.

Butler, R.N. *Why Survive? Being Old In America*. New York: Harper & Row, 1975.

Butler, R.N. and Lewis, M.I. *Sex After Sixty: A Guide For Men And Women In Their Later Years*. New York: Harper & Row, 1976.

Butler, R.N. et al. Self-Care, Self-Help, and The Elderly. *International Journal Of Aging and Human Development*, 1979-1980, *10* (1), 96-97.

Butler, R. and Gleason, H. *Productive Aging*. New York: Springer Publishers, 1985.

Collins, A.H. and Pancoast, D.L. *Natural Helping Networks: A Strategy For Prevention*. Washington, DC: National Association of Social Worker, 1976.

Curry, R.C. *Training for Trainers: Serving The Elderly, The Technique, Part 6*. Durham, NH: New England Gerontology Center, 1980.

DeRopp, Robert S. *Man Against Aging*. New York: St. Martin's Press.

Eizdorfer, C. and Lawton, M.P. (Eds.). *The Psychology of Adult Development and Aging*. Chicago: American Psychological Association.

Erikson, E. ed. *Adulthood*. New York: W.W. Norton and Co., 1978.

Ernst, M. Sensory Impairment Increases With Age: Societal Attitudes Hinder Compensation. *Generations*, 1980, *5*.

Evans, L.K. Maintaining Social Interaction As Health Promotion In The Elderly. *Journal Of Gerontological Nursing*, 1979, *5*(2): 19-21.

FallCreek, S. and Mettler, M. *A Healthy Old Age: A Sourcebook For Health Promotion With Older Adults*. Seattle, WA: Center for Social Welfare Research, School of Social Work, University of Washington, 1984.

FallCreek, S. and Stam, S.B. (Eds.). *The Wallingford Wellness Project—An Innovative Health Promotion Program With The Elderly*. Seattle, WA: Center for Social Welfare Research, School of Social Work, University of Washington, 1981.

Feil, B. *V/F Validation, The Feil Method, How To Help Disoriented Old-Old*. Ohio: Edward Feil Production, 1982.

Freedman, J.L. and Ahronheim, J.C. Nutritional Needs of the Elderly: Debate and Recommendations. *Geriatrics, 40*: 45-59. 1985.

Hearing Loss In The Aged. Anon., Environmental Communications Intervention For The Aging Project, Murray State University, KY.

Helfan, Arthur E., DPM. Common Food Complications In The Elderly Diabetic. *Journal Of The American Podiatry Association*, June 1977, *67*(6): 406-408.

Hess, B.B. Self-Help Among The Aged. *Social Policy*, 1976, *7*: 55.

Kaminsky, M. *The Uses of Reminiscence: New Ways of Working With Older Adults*. New York: The Haworth Press, 1984.

Knopf, Olga. *Successful Aging*. New York: Viking Press, 1974.

Kubler-Ross, E. *Death: The Final Stage of Growth*. Englewood Cliffs, NJ: Prentice-Hall, Inc., 1975.

Kubler-Ross, E. *On Death and Dying*. New York: Macmillan, 1968.

Lambing, M. L. et al. *Idea Exchange: Preventing Misuse of Drugs By Elders*. Paper presented at Western Gerontological Society Annual Meeting, Anaheim, CA, March 1980.

Luce, G. *Your Second Life*. New York: Delacorte Press, Seymour Lawrence, 1979.

Mezey, M., Rauckhorst, L., and Stokes, S. *Health Assessment Of The Older Individual*. New York: Springer Publishing Co., 1980.

Miller, I. and Solomon, R. The Development of Group Services for the Elderly. *Journal of Gerontological Social Work*, 1980, 2(3): 247-248.

Monk, A. Family Supports In Old Age. *Social Work*, 1979, 24(6): 534.

Mookerjee, B.K., Lianos, E.A., Herman, T.S. and Bentzel, C.J. Hypertension In The Elderly. *Geriatric Medicine*, 1982, 9: 94-103.

Oyer, H. and Oyer, E.J. (Eds.). *Aging And Communication*. Baltimore: University Park Press, 1976.

Safe Use Of Medicine By Older People. *Age Page*. National Institute on Aging, November 1980.

Saxon, Sue V. and Etten, Mary Jean. *Physical Change And Aging: A Guide For The Helping Professions*. New York: The Tiresias Press, 1978.

Schrock, Miriam. *Holistic Assessment Of The Healthy Aged*. New York: John Wiley and Sons, 1980.

Schuckit, M.A. Sensitivity Complicates Elders' Substance Abuse. *Generations*, 1980, 5(2), 8-9:36.

Shanas, E. The Family As A Social Support System In Old Age. *The Gerontologist*, 1979, 19(2), 169.

Shore, H. Designing A Training Program For Understanding Sensory Losses In Aging. *The Gerontologist*, 1976, 16.

Smith, Bert K. *Aging In America*. Boston: Beacon Press, 1973.

Statement On Hypertension In The Elderly. National High Blood Pressure Information Center, National Institutes of Health, Bethesda, MD, 1980. In I. Rossman (Ed.) *Clinical Geriatrics* (2nd ed.). Philadelphia, PA: J. B. Lippincott Company, 1979.

Steward, D. (Ed.). *A Fine Age, Creativity As A Key To Successful Aging*. Arkansas: August House Publishers, 1984.

The Aging Eyes: Facts on Eye Care For Older Persons. Anon., pamphlet, National Society To Prevent Blindness, 1979.

Tift, *Meeting The Psychosocial Needs Of The Older Person*. Minneapolis, MN: The Program on Aging of Augsburg College, 1977.

Toseland, R.W. et al. A Community Outreach Program For Socially Isolated Older Persons. *Journal of Gerontological Social Work*, 1979, *1*(3).

Tournier, Paul. *Learn To Grow Old*. New York: Harper and Row, 1972.

Tress, J. Family Support Systems For The Aged: Some Social and Demographic Considerations. *The Gerontologist*, 1977, *17*(6): 487-488.

Vestal, R.E. *Drugs And The Elderly*. National Institute On Aging, Science Write Seminar Series, pub. # 79-1449, July 1979.

Warner-Reitz, Anne. *Healthy Lifestyles for Seniors, A Program Development Manual*. Meals for Millions/Freedom from Hunger Foundation, 1981. Copies of this book can be ordered from the publisher: Meals for Millions/Freedom from Hunger Foundation 815 Second Avenue, Suite 501, New York, NY 10017.

Watkins, D.M. Nutrition For Aging and the Aged. In Goodhart, R.D., (Ed.) *Modern Nutrition In Health and Disease* (p. 781). Philadelphia: Lea and Febiger, 1980.

Winter, Ruth. *Ageless Aging: How Science Is Winning The Battle To Help You Extend Your Healthy And Productive Years*. New York: Crown Publishers, 1973.

Young, E.A. Nutrition, Aging and the Aged. *Medical Clinics of North America*, 1983, 67: 295-313.

HEALTH CARE

Arthritis, The Basic Facts. Anon., The Arthritis Foundation, 1978.

Bauman, E., Brint, A.I., Piper, L., and Wright, P.A. *The Holistic Health Handbook*. Berkeley: Berkeley And/Or Press, 1978.

Biermann, June and Toohey, Barbara. *The Diabetics Total Health Book*. Los Angeles: J.P. Tarcher, Inc., 1980.

Brody, J. *Jane Brody's Nutrition Book*. New York: Bantam Books, 1982.

Carr, R. *Arthritis: Relief Beyond Drugs*. New York: Harper and Row, 1981.

Cousins, N. *Anatomy Of An Illness*. New York: W.W. Norton & Co., 1979.

Cousins, N. *Human Options*. New York: W.W. Norton & Co., 1981.

Dass, Ram and Levine, S. *Grist For The Mill*. Santa Cruz, CA: Unity Press, 1976.

Dass, Ram. *Journey of Awakening*. New York: Bantam Books, 1978.

Dass, Ram and Gorman, P. *How Can I Help?* New York: Alfred A. Knopf, 1986.

Dolan, Y.M. *A Path With A Heart, Eriksonian Utilization With Resistant and Chronic Clients*. New York: Brunner/Mazel, Inc., 1985.

Dychtwald, K. *Body-Mind*. New York: Jove Publishers, 1977.

Eliot, R. and Breo, D. *Is It Worth Dying For?* New York: Bantam Books, 1984.

Frank, A. and Frank, S. *The People's Handbook Of Medical Care*. New York: Vintage Books, 1972.

Good Nutrition In Later Years. Pamphlet by Channing L. Bete Co., Inc., South Deerfield, MA 01373.

Graubarth-Syzller, B., Padgett, J., and Weiss, J.C. *Discovering Wellness In A Nursing Home*. New Orleans, LA: Longevity Therapy, Inc., 1987.

Hamilton, E.M., Whitney, E., and Sizer, F.S. *Nutrition, Concepts and Controversies* (3rd ed.). New York: West Publishing Company.

Knopp, S. *Back To One, A Practical Guide For Psychotherapists*. California: Science and Behavior Books, Inc., 1977.

Lorig, K. and Fries, J. F. *The Arthritis Helpbook: What You Can Do For Your Arthritis*. Reading, MA: Addison-Wesley Pub. Co., 1980.

Lowe, J. *Viewpoint: Toward A Healthier America*. W.K. Kellogg Foundation, January 1980.

Memmler, R. and Rada, R. *The Human Body In Health And Disease*. Philadelphia: J. B. Lippincott Company, 1970.

Patient Instructions For The Care Of The Diabetic Foot. San Francisco, CA: Handout, California College of Podiatric Medicine.

Pelletier, K. *Mind As Healer, Mind As Slayer*. New York: Delacorte Press, Seymour Lawrence, 1977.

Peterson, B. H., Kennedy, B. J., Butler, R.N., and Gastel, B. *Aging and Cancer Management*. New York: American Cancer Society, 1979.

Simon, S.B., Howe, L.W., Kirschenbaum, H. *Values Clarification*. New York: Hart Publishing Co., 1972.

Simonton, O.C., Matthews-Simonton, S., and Creyton, J. *Getting Well Again: A Step-By-Step, Self-Help Guide To Overcoming Cancer For Patients And Their Families*. Los Angeles: J.P. Farcher, 1978.

Six Steps To Help Control Your Arthritis Symptoms. Pamphlet from the McNeil Pharmaceutical, McNeilab, Inc., Spring House, PA, 1977.

Skeist, R.J. *To Your Good Health*. Chicago Review Press, 1980. Wallingford Wellness Project Staff Team. Health Promotion Educational Materials. Seattle, WA: Center for Social Welfare Research, School of Social Work, University of Washington, 1982.

Steinmetz, J., Blankenship, J., Brown, L., Hall, D., and Miller, G. *Managing Stress: Before It Manages You*. Palo Alto, CA: Bull Publishing Company, 1980.

Successful Diet and Exercise Therapy Conducted In Vermont For Diabesity. *Journal of the American Medical Association*, 1980: *243*.

The American National Red Cross, *Standard First Aid And Personal Safety*, Garden City, NY: Doubleday & Co. Inc., 1973.

The Calcium Cookbook. Pamphlet by the AARP, Health Advocacy Services, Program Department, 1909 K Street, N.W., Washington, DC 20049.

Tubesing, D.A. *Kicking Your Stress Habits, Y's Way To Stress Management*. Illinois: Whole Person Associates, Inc., 1981. Distributed by The YMCA Program Store, P.O. Box 5077, Champaign, IL 61820, (312) 351-5077.

U.S. Department of Agriculture, U.S. Department of Health, Education and Welfare. *Nutrition and Your Health: Dietary Guidelines For Americans*, 1981.

Water, How 8 Glasses A Day Keep Fat Away, by Donald S. Robertson, MD, MSc., *McCall's Magazine*, January 1986.

Weiss, Jules C. *Expressive Therapy With Elders and the Disabled: Touching The Heart Of Life*. New York: The Haworth Press, 1984.

Why Risk Heart Attack? Pamphlet, American Heart Association.

Williams, Marcia. *No Salt, No Sugar, No Fat, No Apologies Cookbook*. Crossing Press, 1986.

Index